Modern Cosmology

Modern Cosmology

D. W. SCIAMA

Fellow of All Souls College, Oxford

CAMBRIDGE UNIVERSITY PRESS

CAMBRIDGE

LONDON · NEW YORK · MELBOURNE

CAMBRIDGE UNIVERSITY PRESS
Cambridge, New York, Melbourne, Madrid, Cape Town, Singapore,
São Paulo, Delhi, Dubai, Tokyo, Mexico City

Cambridge University Press
The Edinburgh Building, Cambridge CB2 8RU, UK

Published in the United States of America by Cambridge University Press, New York

www.cambridge.org
Information on this title: www.cambridge.org/9780521287210

First published 1971
Reprinted with corrections 1972
Reprinted 1973, 1975
Re-issued 2010

A catalogue record for this publication is available from the British Library

Library of Congress Catalogue Card Number: 73–142961

ISBN 978-0-521-08069-9 Hardback
ISBN 978-0-521-28721-0 Paperback

Contents

To Lydia

Preface

It has now become clear that the exploration of the Universe, as conducted by physicists, astronomers and cosmologists, is one of the greatest intellectual adventures of the mid-twentieth century. It is no exaggeration to say that their achievements, especially in the last few years, constitute a revolution in our knowledge and understanding of the Universe without parallel in the whole recorded history of mankind. In this book I have attempted to tell the story of this revolution in a manner intelligible to readers with only a modest knowledge of mathematics and physics. In selecting the topics to emphasise I have been guided by one consideration alone, that the ultimate aim of this work is to elucidate the structure and history of the Universe as a whole. This is by no means the main aim of all the scientists concerned and in presenting my synthesis of their work I freely admit that I am expressing my own cosmological yearnings. However, I do believe that in retrospect the greatest achievement of our age in astronomy will be seen to be the new insight gained into the workings of our Universe on the largest possible scale.

In accordance with this aim I have planned the book as follows. The first few chapters essentially describe the main contents of the Universe as they are known today – stars, galaxies, radio galaxies and quasi-stellar objects (QSOs). No discussion of these objects, especially of the QSOs, would make much sense without detailed reference to that key feature, the large red shifts of the lines in their optical spectra. These red shifts lead to the idea that the Universe is expanding and accordingly the main features of this expansion are described at the same time as the contents of the Universe. Theories of the expansion are discussed in chapter 8.

This is the longest, most mathematical and most difficult chapter in the book. I could not resist writing it in this form because I believe that anything simpler would not do justice to my theme. I have, however, summarised the main theoretical results at the end of the chapter, beginning on p. 117. For many purposes this summary should suffice.

The theoretical discussion throws up the possibility that intergalactic space may contain a significant quantity of matter and be the seat of important physical activity. These questions are discussed in the next two chapters. Finally we consider that most remarkable of discoveries, the 3 °K cosmical microwave radiation. The significance of this radiation, its relation to the hot big bang and the helium problem, to cosmical high energy processes and to questions of isotropy, occupy us in the last six chapters. Most of this work is less than four years old, and is clearly in its infancy. Here we stand at the threshold of the cosmology of the future.

In writing this book I have been very conscious of what I owe to my students, colleagues and predecessors. I have borrowed freely from many of their articles and much of their conversation, and if I record here the names of E. M. Burbidge, A. Sandage and R. J. Tayler it is because my borrowing from their writings has been particularly extensive. I am grateful to T. Gold for his comments on the manuscript, to G. G. Pooley for his help with several of the illustrations, to M. P. Coles and other members of the Cambridge University Press for their work on the text and the diagrams, to Mrs M. D. Anderson for compiling the indexes, and to Faber and Faber for giving me permission to use six diagrams and about three pages of text from my book *The Unity of the Universe*. Above all I owe a great debt of gratitude to my student and colleague M. J. Rees whose physical insight has helped to guide me through the unfamiliar territory of the new Universe.

D. W. SCIAMA

Venice – Cambridge 1967–9

Note added in 1975

Six years have passed since this book was first completed, and two years since it was reprinted with an additional note. These last two years have seen one development whose potential importance for cosmology is comparable to the recent ones already recorded, and that is the discovery of deuterium outside the solar system. It is possible that this deuterium was made in the hot big bang (see fig. 48), in which case its observed abundance would be a much more sensitive indicator than that of helium for the present mean density of the Universe, and so for its future evolution; that is, for whether the expansion will continue indefinitely, or whether the Universe will re-contract into another big bang.

Prior to 1973 deuterium was known to exist in the oceans and in meteorites, but had never been detected outside the Earth. In that year no less than six independent investigations led to the discovery of extraterrestrial deuterium, at radio, visible and ultra-violet wavelengths, in a variety of objects including the planet Jupiter, the Orion nebula and interstellar clouds. In all these cases except one the deuterium was chemically associated with other atoms, so that a reliable cosmic abundance of deuterium relative to hydrogen could not be obtained, owing to the uncertainty in the correction for chemical fractionation. Fortunately the exceptional case does lead to a good value for the deuterium abundance. This was derived from the observation by J. B. Rogerson and D. G. York of the Lyman lines of atomic deuterium in the ultra-violet absorption spectrum of the star β Centauri, obtained from above the Earth's atmosphere by the satellite Copernicus. The resulting D/H ratio by number is $1.4 \pm 0.2 \times 10^{-5}$, a value which is consistent with the more uncertain estimates made from the molecular observations.

This result is of capital importance for cosmology. The reason for this stems from the difficulty of making deuterium by stellar processes occurring in the Galaxy. Indeed stars tend to destroy deuterium by nuclear burning rather than manufacturing it. Attempts have been made to explain the production of deuterium by appealing to specialised processes occurring in the envelopes of supernovae or in an earlier generation of supermassive stars. Such attempts have not yet been successful and, although better models may be proposed in future, current thinking amongst astronomers suggests that most of the observed deuterium was probably formed by thermonuclear reactions occurring in the hot big bang. Reference to fig. 48 then shows that to account for the observed abundance of deuterium, on the assumption that the universe is filled with a 2.7 °K black body radiation field, requires the present mean density of the Universe to be about 10^{-30} g cm^{-3}. As explained on pages 110 and 118, *this corresponds to a low-density Universe which is destined to expand forever.*

The importance of this conclusion for cosmology hardly needs underlining. It does depend, however, on accepting that relatively little deuterium is manufactured in the Galaxy. This is difficult to prove directly, since one can hardly enumerate all conceivable specialised processes, and then show that they are all inadequate. Strong evidence would come from the establishment of great *uniformity* in the distribution of deuterium in the Galaxy. On the other hand the absence of uniformity would not rule out cosmological production, since subsidiary processes may have occurred later to change the abundance of deuterium in particular locations in the Galaxy. For instance the material one is observing may have passed through regions in stars where deuterium is destroyed, later to be expelled back into the interstellar medium. Or it may have recently been accreted by the Galaxy from intergalactic space, where it would more nearly reflect the primeval abundance. Indeed the best evidence of all for cosmological production would be the direct detection of *intergalactic* deuterium whose abundance relative to hydrogen would be decisive. However, at the time of writing even intergalactic hydrogen itself has not yet been detected.

Similar considerations apply to some of the isotopes of other light elements, namely lithium, beryllium and boron, which also tend to be destroyed in stars. Here there is a further complication, because additional likely sources of some of these isotopes are interstellar carbon, nitrogen and oxygen which, under bombardment by cosmic ray protons and α-particles, break up into the lighter elements. Indeed if the cosmic ray flux has been essentially constant throughout the life of the Galaxy (10^{10} years), one can account for the observed abundances of ^{6}Li, ^{9}Be, ^{10}B and ^{11}B entirely in terms of cosmic ray induced processes. On the other hand one cannot account in this way for the observed abundances of D, ^{3}He, ^{4}He or ^{7}Li. It is therefore remarkable that, conversely, one can (more or less) account for the D, ^{3}He, ^{4}He and ^{7}Li by cosmological production with a unique assumed (low) density for the Universe, but not the ^{6}Li, ^{9}Be, ^{10}B and ^{11}B. While many of the details still remain to be cleared up, this beautiful fitting together like a jig-saw puzzle of the cosmic ray and the cosmological processes strongly suggests that we are here on the right track, and therefore that the Universe is destined to expand forever. The reader interested in pursuing these questions is strongly recommended to study the excellent review article by Hubert Reeves listed in the bibliography at the end of this book.

The note added to the 1973 printing already recorded the other important advance of recent years, namely the observational discovery that rich clusters of galaxies, such as those in Virgo, Coma and Perseus, emit a copious flux of X-rays, with a power in the range 10^{43}–10^{44} ergs per second. We owe this and many other discoveries of importance in X-ray astronomy (such as the possible discovery of a black hole in Cygnus X-1) to the first X-ray satellite UHURU. Other X-ray satellites are now in operation and further ones are planned, so there are still important discoveries likely to be made. For cosmology the most important step would probably be to determine the origin of the diffuse X-ray background, which, as discussed in chapters 10 and 15, probably arises from sources distributed throughout the Universe.

The extension of this background into the γ-ray region at present contains greater observational uncertainties, but is also of cosmological importance.

Another development of interest to cosmology is the growing observational evidence that clusters of galaxies may contain a significant quantity of intergalactic gas, with a likely density in the range 10^{-28}–10^{-27} g cm^{-3}. As mentioned on pages 40, 41 and 129, the existence of such a gas would be relevant to the question of whether clusters of galaxies are gravitationally bound. The new evidence comes from the X-ray observations referred to in the previous paragraph, and from observations of radio galaxies which suggest that there may be an intergalactic wind blowing on them. If the X-rays are emitted by hot gas (and not by the inverse Compton process (page 186), which is still a possibility) then analysis shows that for a given cluster of galaxies the *same* density of gas would account simultaneously for the X-ray and radio observations. However, this density would *not* be sufficient to bind the cluster gravitationally, so the well-known virial discrepancy (pp. 40–1) remains. On the other hand, the suspicion is growing amongst astronomers, independently of the virial problem, that the masses of galaxies may have been underestimated by a substantial factor, so that the resolution of the problem may lie in that direction.

For the rest it has been a matter mainly of accumulating further data of a kind already described in the book. We may mention the following points:

(i) Independent attempts to detect the intense flux of gravitational waves reported by J. Weber have not been successful (pp. 33, 34).

(ii) G. Burbidge no longer believes that the absorption red shift of 1.95 has a special significance (p. 77).

(iii) Two 18th magnitude QSOs have been discovered with red shifts exceeding 3. They are OH 471, whose red shift of 3.40 was obtained by R. F. Carswell and P. A. Strittmatter, and OQ 172 whose red shift of 3.53 was obtained by E. J. Wampler, L. B. Robinson, J. A. Baldwin and E. M. Burbidge. Owing to the large red shifts, the colours of these objects are not typical for QSOs and they were discovered entirely on the basis of very accurate radio positions (better than one second of arc). Now that radio positions of this accuracy are generally obtainable, it has become possible to make optical identifications on position-coincidence alone, without having to appeal also to some special optical property of the candidate object. This development will usher in a new era of statistical studies on radio sources which is likely to have profound implications for cosmology (cf. chapters 6 and 7). However, while we may now expect the discovery of other QSOs of large red shift but atypical colours, it is likely that QSOs of red shift exceeding 2.5 are still relatively rare (p. 79).

(iv) H. C. Arp and others have claimed that there exist further physical associations of objects with very different red shifts. The status of this claim is still under discussion (p. 80).

(v) Galaxy-like images have been discovered by J. Kristian around those QSOs where it would be possible to observe them. This result supports the cosmological hypothesis for the origin of QSO red shifts (p. 80).

(vi) Five QSOs have been found to have the same red shifts as the clusters of galaxies surrounding them, which further strengthens the case for the cosmological hypothesis (p. 82).

(vii) Recent data have further weakened the case for a non-random distribution of QSOs with large red shifts (p. 94).

(viii) Recent measurements of the helium abundance in the Galaxy support the hypothesis that this abundance is uniform and close to the value derived from the canonical hot big bang (chapters 11, 13).

(ix) Further measurements of the cosmic background below 1 millimetre are in conflict with the previously reported excess, which has itself now been withdrawn. The new measurements are in agreement with a black body spectrum at 2.7 °K. In particular, a recent balloon flight by E. I. Robson, D. G. Vickers, J. S. Huizinga, J. E. Beckman and P. E. Clegg of Queen Mary College, London University has shown for the first time that the flux does really fall at wavelengths shorter than the Planck peak for a temperature of 2.7 °K (~2 mm). A further study of this wavelength region by balloon and satellite should be very rewarding (p. 180).

(x)   Recent observations at 2.64 mm rule out the possibility that the rotational excitation of interstellar CN molecules arises from collisions with material particles. In addition radio observations of interstellar formaldehyde lead to an upper limit of 2.9 °K for the radiation temperature at 2 mm (p. 182).

(xi)   Absorption from the second rotational state of CN has now been observed, leading to an excitation temperature of 2.9 °K at 1.32 mm (p. 184).

(xii)   According to present evidence the cosmic ray spectrum does not flatten out at 10^{18} eV (fig. 51 and p. 191).

(xiii)   It now seems doubtful whether the cosmic ray spectrum in the range 10^{21}–10^{28} eV will be measured in the near future (p. 191).

(xiv)   Experiments now being conducted may lead to a determination of our motion through the microwave background with a precision of about 30 km s^{-1} (p. 199).

On the theoretical side there have been studies of the original singularity in the Universe (p. 127), of pair creation close to the singularity, of anisotropic universes and the dissipation of anisotropy (p. 200), and of the formation of galaxies and the black body radiation field. These studies are of great interest but have not yet reached the stage where they should be described in a general book of the present kind.

I am grateful to Professor H. J. Rood for pointing out many misprints in the first printing and to Professor J. G. Wilson for his advice about the cosmic ray spectrum.

D. W. SCIAMA

Oxford January 1975

1 The physics of the stars

INTRODUCTION

There are many types of star amongst the 10^{11} or so that make up our Galaxy. By far the most common type are the so-called main sequence stars, which have a well-defined relation between their power output and their surface temperature. Astronomers show this relation on the so-called Hertzsprung–Russell diagram (fig. 1). Fortunately our own Sun is itself a main sequence star, and it can, of course, be studied in far greater detail than any other star. The physics of the Sun, then, is essentially the physics of the great majority of stars in our Galaxy and, very probably, in most other galaxies as well. We shall therefore begin our study of the stars by considering in detail the properties and structure of the Sun. The less common types of star are, none the less, of great interest and become of central importance when we come to study the *evolution* of the stars (p. 8).

THE PHYSICS OF THE SUN

The main observed properties of the Sun are given in round figures in table 1.

TABLE 1 *Observed properties of the Sun*

Distance	1.5×10^{13} cm
Angular diameter	$0.5°$
Apparent luminosity	1.5×10^6 erg s^{-1} cm^{-2}
Surface temperature	$5000°$
Age	$> 3 \times 10^{16}$ s
Orbital period of the Earth	3×10^7 s

The distance of the Sun can be determined directly by radar, that is, from the time taken by a radio wave to travel to the Sun and to be reflected back to Earth. The apparent luminosity is the rate at which radiant energy from the Sun arrives at unit area of the Earth's surface and is directly measurable. By contrast the Sun's surface temperature must be estimated indirectly, for example, from its colour (the hotter an incandescent body the bluer its

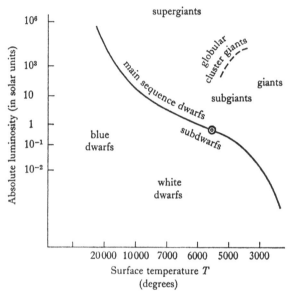

Fig. 1. The Hertzsprung–Russell diagram, which gives the relation between the absolute luminosity L and the surface temperature T of the stars in the Galaxy. Most stars lie close to the curved line (the main sequence). The non-uniform temperature scale corresponds to a uniform variation of the directly measured quantity – the colour-index.

radiation). Finally the lower limit on its age is determined from geological considerations concerning the temperature of the Earth in the distant past.

We include in table 1 the period of the Earth in its orbit around the Sun because we can use it to determine the mass of the Sun. All we need in addition are Newton's inverse square law of gravitation and the distance of the Earth from the Sun. We can also deduce a number of other properties of the Sun from simple geometry and the inverse square law of apparent luminosity. These derived properties are listed in table 2.

TABLE 2 *Derived properties of the Sun*

Radius	7×10^{10} cm
Mass	2×10^{33} g
Mean density	1.4 g cm^{-3}
Absolute luminosity	4×10^{33} erg s^{-1}

We can now use simple physical arguments to infer some further properties of the Sun. The most immediate inference arises from the fact that the Sun has changed little over the last 3×10^{16} seconds at least. If the Sun were unsupported against its own gravity it would collapse in a far shorter time. This collapse time can be estimated from gravitational theory or dimensional considerations to be about $(G\bar{\rho})^{-\frac{1}{2}}$, where G is the Newtonian constant of gravitation and $\bar{\rho}$ is the mean density. With $G \sim 6.7 \times 10^{-8}$ g^{-1} cm^3 s^{-2} and $\bar{\rho} \sim 1.4$ g cm^{-3}, we obtain for the collapse time about 3000 seconds or 1 hour. Clearly the balance between gravitation and the supporting forces must be a very exact one to hold the Sun stationary for at least 3×10^{16} seconds or 10^9 years.

There can be no question of these forces being solid body forces such as those that support the Earth. For the Sun is 330 000 times more massive than the Earth, and would easily crush any cold material into the white dwarf condition described on p. 9. What keeps the Sun distended is the heat motion of its constituent particles, or, more exactly, the thermal pressure gradient existing between its centre and its surface. We can estimate the mean temperature through the body of the Sun by requiring that the thermal kinetic energy of the Sun be equal to its gravitational potential energy. This potential energy we can estimate roughly by assuming that the Sun is a uniform sphere of mass $M_\odot$ and radius R. The potential energy is then $\frac{3}{5}GM_\odot^2/R \sim 10^{48}$ ergs, which corresponds to $\frac{3}{5}GM_\odot/R$ or 5×10^{14} ergs per gramme of the Sun's material. If the atoms in the Sun have a mean thermal speed v, then the thermal kinetic energy per gramme is $\frac{1}{2}v^2$. Equating this to 5×10^{14} we find for v the value 300 km s^{-1}. If the gas is in the form of hydrogen the corresponding mean temperature is 5×10^6 °K. This is a very important result since hydrogen of mean density ~ 1 g cm^{-3} at the high temperature of 5×10^6 °K will be nearly completely ionised, that is, the electrons will be set free from the protons by virtue of the vigorous collisions that occur at 300 km s^{-1}. This greatly simplifies the study of the Sun's structure because it means that even at a density of 1.4 g cm^{-3} *the material of the Sun behaves like a perfect gas* (the spacing between its constituent particles is about 10^{-8} cm which is comparable with the radius of a hydrogen atom but very much greater than the size of a proton or an electron). Had the material of the Sun been solid or liquid its structure would have been much more complicated.

The Sun is thus a ball of gas with a central temperature greater than 5×10^6 °K and a surface temperature of 5000 °K. The corresponding mean temperature gradient is about 10^{-4} °K cm^{-1}. The heat radiation at the centre of the Sun is not directly visible because on its way to us it is scattered and absorbed. The simplest of these processes is scattering by the free electrons in the Sun – the so-called Thomson effect. The resulting mean free path λ of the radiation for this process alone is given by

$$\lambda \sim \frac{10^{24}}{n} \text{ cm},$$

where n is the number of free electrons per cubic centimetre. In the Sun $n \sim 10^{24}$ and so $\lambda \sim 1$ cm. Accordingly the sun is highly opaque. Moreover the temperature changes by only 10^{-4} °K over one mean free path, so that no difficulty arises in speaking about a temperature at each part of the Sun. Nevertheless the temperature gradient must not be forgotten, for it is needed to drive the energy flux against the opacity of the Sun's material.*

This last point enables us to make a simple rough theoretical estimate of the Sun's luminosity. The energy density in a radiation field of mean temperature $\bar{T}$ is, by Stefan's law, $a\bar{T}^4$, where a is

---

* In a complete discussion the transport of energy by conduction and convection would also have to be considered, but in fact over most of the Sun radiative transport dominates.

Stefan's constant (7.59×10^{-15} erg cm^{-3} deg^{-4}). Thus the energy flux per square centimetre is roughly

$$\lambda c \frac{\mathrm{d}}{\mathrm{d}r} (a \bar{T}^4),$$

where c is the velocity of light and r is a radial co-ordinate. Assuming the Sun to be in a steady state, this is also the rate at which energy leaves a square centimetre of the surface, that is, it is essentially the Sun's absolute luminosity. From our estimates of λ and the temperature gradient we in fact obtain a fairly good value for the Sun's luminosity, but the accuracy of this estimate is, of course, limited by the uncertainty in $\bar{T}$ which occurs to a relatively high power.

So far we have been considering solar quantities only. In fact we have deduced everything from just the mass and radius of the Sun. The question then arises: what is the luminosity of a star whose mass M and radius R are different from those of the Sun? On the basis of our simple estimates we have

$$\bar{T} \propto \frac{M}{R},$$

$$\lambda \propto \frac{1}{n} \propto \frac{R^3}{M}.$$

Hence the energy flux per square centimetre is roughly proportional to

$$\frac{R^3}{M} \times \frac{M^4}{R^5}$$

$\left(\text{since} \dfrac{\mathrm{d}\bar{T}}{\mathrm{d}r} \text{ is roughly } \dfrac{\bar{T}}{R}\right),$

or to

$$\frac{M^3}{R^2}.$$

Thus the luminosity L over the whole surface satisfies the relation

$$L \sim M^3.$$

This is a simple example of Eddington's famous mass–luminosity relation. In our derivation the radius completely cancels out, but in a more detailed discussion it is found that the luminosity does depend on the radius, although weakly. These calculations are fairly well supported by observations of the masses and luminosities of main sequence stars (the masses being determined from the orbital characteristics of binary stars). The most important consequence of the mass–luminosity relation is that stars more massive than the Sun are much more luminous, and are therefore using up their energy sources much more rapidly.

This brings us to the question of the source of energy for a typical star. Indeed the reader may have been puzzled at our ability to find out so much about a star without specifying either the mechanism or the rate of energy generation. The reason is that our assumption that the Sun is stable for at least 10^9 years puts constraints on the mechanism of energy generation, which of course the actual mechanism must in fact satisfy. These constraints arise as follows. In the first place, we note that the heat content of the Sun is 10^{48} ergs, which is radiated away at the rate of 4×10^{33} erg s^{-1}. Thus at any one time the Sun contains enough heat to radiate for 2.5×10^{14} s or about 10^7 years. But the Sun has been more or less unchanged for at least 10^9 years. Thus it must be generating energy continuously at the same rate L as it radiates.

Now whatever the mechanism of energy generation its rate will probably depend on $\bar{\rho}$ and $\bar{T}$ which are determined by M and R. Thus at first sight there is no reason why the energy generation rate should in fact be L, which as we have seen is also determined by M and R, but presumably not by the same relation. Suppose for instance that the energy generation rate were too low. The star would then cool down and so would no longer be able to support itself against gravity. Accordingly, the star would contract. This contraction would in turn lead to an *increase* in the temperature of the star, and it is likely that this would increase the rate of energy generation. Clearly the star is seeking a stable state and it would find one if the energy generation rate depends sufficiently strongly on the temperature. For in that case the star would contract and heat up until energy is being generated internally at just the rate it is being lost at the surface. Thus the radius of the steady star would be determined by its mass and the mechanism of energy generation.

What is this mechanism? In the Sun it must generate energy at the rate of 2 erg s^{-1} g^{-1}, so that in 10^9 years (3×10^{16} s) it generates

6×10^{16} erg g^{-1}. But the gravitational potential energy of the Sun is only 5×10^{14} erg g^{-1}. In other words the potential energy released during the original contraction of the Sun after its formation (the Kelvin–Helmholtz energy) would keep the Sun alight for only about 10^{7} years. There must then be some other mechanism of energy generation. It cannot involve chemical energy with its typical release of only about 10^{13} erg g^{-1}. The natural suggestion then is nuclear energy. This is in fact numerically quite reasonable. The binding energy per unit mass of an α-particle (helium nucleus) is about 6×10^{18} erg g^{-1}. Thus in 10^{9} years it would require only about 1 per cent of solar hydrogen to be converted into helium to account for the Sun's luminosity. The Sun is actually believed to be about five times older than this, which would still leave a comfortable margin.

It is a fascinating problem of nuclear physics to work out whether reactions will occur at temperatures somewhat in excess of 5×10^{6} °K to convert hydrogen into helium, or rather, protons into α-particles. By the standards of modern nuclear physics the kinetic energy of the protons is very low (about 500 electron volts; 1 electron volt being about 10^{-12} ergs) but we now know that such nuclear reactions can occur. They were first worked out by C. F. von Weizsäcker and H. A. Bethe in 1938–9,* and have since been re-investigated by many other physicists. The γ-rays produced in these reactions are degraded into X-rays and then into visible light before they reach us because, as we have seen, the Sun is opaque to these rays. But these reactions also produce particles called neutrinos which, like γ-rays and X-rays, move with the speed of light, but interact very *weakly* with ordinary matter. The Sun is almost completely transparent to these neutrinos, and a very bold attempt to detect them is now being made by the American nuclear physicist, R. Davis. If successful this experiment would correspond to 'seeing' directly into the Sun's central regions where the temperatures are highest and most of the thermonuclear reactions are occurring. The present situation is an intriguing one. Davis has already been able to place an upper limit on the solar neutrino flux at the Earth that is significantly less than the originally

* Professor Bethe was awarded the Nobel Prize in 1967 for this work.

predicted value. This has greatly perturbed the theorists, and has suggested that we understand less about the structure of even main sequence stars than was formerly thought.

STELLAR EVOLUTION

What happens to a star which has converted into helium all the hydrogen in its central hottest regions? This question is particularly pertinent for stars more massive than the Sun, since they consume their nuclear fuel at a relatively fast rate. As we have seen, when its energy generation rate is too low a star must first cool down and then contract and heat up again. This contracting and heating would continue indefinitely unless a new set of energy-producing nuclear reactions becomes important. Such nuclear reactions do in fact exist, and significant helium burning begins when the central temperature has reached about a hundred million degrees. In this case the end product is mostly carbon. When the helium is itself consumed the star contracts again, but these later stages of evolution are not as well understood as the earlier ones. In many cases it appears that the star cannot continue to find a stable state in a controlled manner, and explosions may occur. The most extreme of these is the supernova phenomenon, in which a large fraction of a star explodes and may temporarily outshine the galaxy that contains it. The famous Crab nebula (plate 1) shows the remnants of such an explosion recorded by Chinese astronomers in A.D. 1054. We shall have much to say about the Crab later since it is the best-studied example of a violent celestial explosion. Such explosions are now known to occur on a larger than stellar scale elsewhere in the Universe, and they form one of the dominant themes of this book.

The final stages of stellar evolution set in when a star contains no more readily combustible nuclear fuel. If the star is much more massive than the sun it continues to contract, and in the absence of explosive processes this contraction continues until the self-gravitation of the star is so strong that its behaviour is no longer described by the Newtonian theory of gravitation. Under these circumstances most physicists believe that the correct theory to use is Einstein's general theory of relativity, although its predictions concerning the behaviour of strong gravitational fields remain to be tested experimentally.* According to this theory an outside observer would find that the star takes an infinite time to collapse to a certain critical radius. This so-called Schwarzschild radius has the value $2GM/c^2$, where c is the velocity of light. For a star 10 times more massive than the sun the Schwarzschild radius would be 30 kilometres, at which point its density would be about 10^{15} g cm^{-3}. This is just the density of matter within an atomic nucleus, and a star in this extraordinary state is called a neutron star. The outside observer would never in fact see the star quite reach its Schwarzschild radius because the gravitational field at the surface would be so intense that the Einstein red shift† would render it too faint to be detected. An observer unfortunate enough to be sitting on the collapsing star would find himself carried through the Schwarzschild radius and on into regions of such high density that present-day physics would probably no longer hold. The correct physics for such regions is quite unknown and remains an outstanding problem for the future.

By contrast, if the star has a mass less than 1.4 solar masses (the Chandrasekhar limit) it can support itself against gravity even when quite cold because the free electrons within it can then exert sufficient outward pressure.‡ Hot stars in this state have been recognised in the sky; they are called white dwarfs. Their radius is about 1 per cent of the solar radius and their mass is about a solar mass. Their density is thus about 10^6 g cm^{-3}, and it is because of this fantastically high density that the Pauli Exclusion Principle becomes important. Matter in this state is said to be quantum mechanically degenerate.

An even more condensed state can exist theoretically, in which a steady star of about a solar mass has a radius of about 10 kilometres. This would be a neutron star of the type we have already mentioned – except that it would not collapse if its mass were less than a few solar masses. In this case the outward pressure would

---

* An elementary account of Einstein's theory is given in my book *The Physical Foundations of General Relativity*.
† The Einstein red shift is discussed in chapter 5 of *The Physical Foundations of General Relativity*.
‡ This is a quantum mechanical effect, arising from the Pauli Exclusion Principle.

arise from the degenerate neutrons, instead of from the degenerate electrons that support a white dwarf. If such neutron stars exist they would be too faint to be detected optically because of their small surface areas. Until recently it was thought that the only way they could be detected (apart from being an unseen component of a double star) was from their X-ray emission, if their

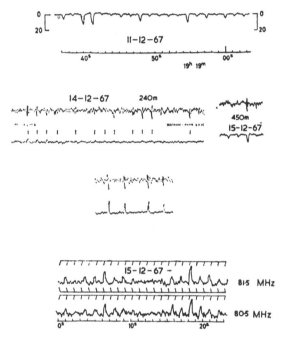

Fig. 2. The first published record of a pulsating radio source (pulsar). (From Hewish *et al.*, *Nature* **217**, 710 (1968).)

surfaces were hot enough ($T \sim 10^7 \, °K$) to emit an appreciable flux of X-rays. However, although some stars are now known to emit X-rays, none of them appeared to be a neutron star.

This was the situation until 24 February 1968. In *Nature* of that date the discovery of pulsars was reported by A. Hewish, S. J. Bell, J. D. H. Pilkington, P. F. Scott and R. A. Collins of the Cavendish Laboratory, Cambridge (see fig. 2). The pulsars are pulsed radio sources. At a given observing frequency their radiation consists of pulses of complex structure with a duration of about 10 milliseconds which repeat very exactly, with a period of 33 milliseconds for the fastest known pulsar and 4 seconds for the slowest. The 33 millisecond pulsar, which is in the Crab nebula, also produces optical and X-ray pulses with the same period. At the time of writing these pulsars have been investigated for only two years, but already much is known about them, and it is clear that their study will revolutionise our understanding of many topics in astronomy. Our immediate concern, however, is their relation to neutron stars. It now seems almost certain that the pulsars are neutron stars, and that the 'clockwork' that imposes the very exact periodicity of the appearance of the pulses is the *rotation* of the neutron star. No other object could produce a period as low as 33 milliseconds that itself increases by as little as the observed 10^{-5} s yr^{-1}. Such a rotating neutron star model for pulsars was first proposed by T. Gold. It is too early to be firm about the details of the model, but its basic idea is almost certainly correct. In particular a rough age estimate from the rate of slowing down of the Crab pulsar agrees with the age of the Crab as determined by the Chinese records. We may therefore conclude that, in all probability, stable neutron stars do actually exist and can be left behind after a supernova explosion.

The various states a star can get into form a complicated scheme and it is convenient to visualise the possibilities by constructing a diagram in which the absolute luminosity of a star is plotted against its surface temperature. This so-called Hertzsprung–Russell (H–R) diagram has already been mentioned and is shown in fig. 1. It will be seen that only portions of this diagram are occupied by actual stars. Typical stars, like the Sun, form the main sequence, and other types of star are shown where they occur on the diagram.

This H–R diagram is also useful for following the evolution of a single star. In fact a main sequence star moves up the main sequence until about 10 per cent of its hydrogen is converted into helium. It then turns off to the right. Since hydrogen burning releases about 6×10^{18} erg g^{-1}, the age of a star at the turn-off point is about $6 \times 10^{18} \, \bar{M}/10 \, \bar{L}$ seconds or about $2 \times 10^{10} \, \bar{M}/\bar{L}$ years, where $\bar{L}$ is the average absolute luminosity of the star. This age can be inferred from observation by studying a cluster of stars in the hope that most of the stars in the cluster were formed at roughly the same time and have a similar composition. The main sequence of such a cluster (fig. 3) would then delineate the range of masses amongst the stars, and the position of the turn-off point would give the age of the cluster. Observed clusters vary in age from

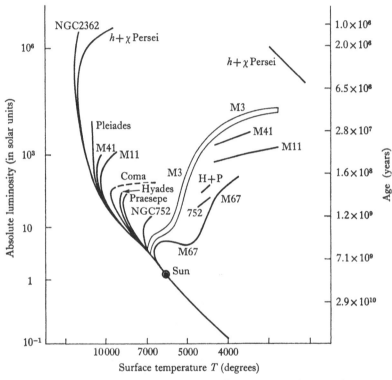

Fig. 3. The Hertzsprung–Russell diagram for ten open clusters of stars and one globular cluster. Ages corresponding to the various turn-off points from the main sequence are given along the right-hand ordinate. (After A. Sandage, *Astrophys. J.* **125**, 435 (1957).)

about 10^6 years at which time the turn-off point has only reached very massive stars, to about 10^{10} years when stars about as massive as the Sun are at the turn-off point. This latter result is of great importance to cosmology since it tells us that stars have existed more or less undisturbed for the last 10^{10} years or so. However, as we shall see, the Universe as a whole has a time-scale of the same order. A little more than 10^{10} years ago, stars could not have existed in the form we see them today. Thus there is a close link between problems of stellar evolution and those of cosmology.

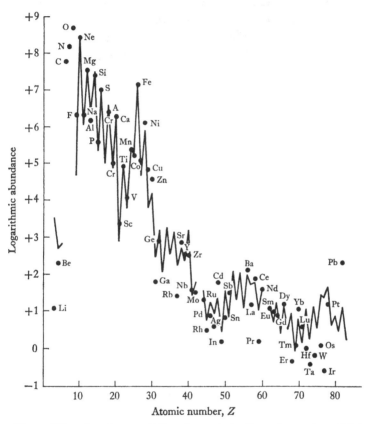

Fig. 4. The abundances of the elements in the solar system. (Galactic abundances are similar, although there are important exceptions.) The dots are derived from the strengths of absorption lines in the solar spectrum. The lines are based mainly on data from the Earth and meteorites. As compiled by H. E. Suess and H. C. Urey.

A further link is provided by our remark that during late stages of stellar evolution explosions may occur in which stellar material is thrown out into the Galaxy. This material would have been processed by the nuclear reactions that occurred both in the stars and during the explosions themselves, and so the material of the Galaxy would become contaminated with elements heavier than hydrogen. This raises the whole question of explaining the observed abundances of the elements in the Galaxy (fig. 4). Were they all formed in hot stars, or did the original composition of the Galaxy include heavy elements? This is clearly a cosmological question, to which we will return in a later chapter (chapter 13).

2 The Milky Way

INTRODUCTION

The Milky Way takes its name from the band of white light which can be seen stretched across the sky on a clear night (plate 2). It was discovered by Galileo that this band of light actually comes from a vast collection of faint stars which the naked eye cannot distinguish. The Sun itself and the nearby stars also belong to this collection, which is held together by the gravitational forces between the stars. The resulting object is called a galaxy (*galaxias* is the Greek word for milk). We now know that the Universe contains a large, perhaps infinite, number of somewhat similar galaxies lying well outside our own Milky Way system and spread out through space at least as far as the most powerful telescopes can penetrate. In this chapter we shall confine our attention to our own Galaxy, leaving for the next chapter a discussion of the external galaxies.

It should be said at once that while most of the mass of our Galaxy is in the form of stars, there is much physical activity occurring in the vast spaces between the stars. Indeed some of the most fascinating phenomena in astrophysics concern the gas, dust, cosmic rays and magnetic fields that lie in interstellar space. These phenomena have been closely studied in recent years both for their own sake and because similar processes underlie the observed behaviour of radio galaxies and quasi-stellar objects. We shall therefore consider these aspects of the structure of our Galaxy after describing its stellar structure.

THE STELLAR STRUCTURE OF THE MILKY WAY

The naked eye can see only 6000 or so of the 10^{11} stars that make up the Milky Way. The structure of this whole system is now reasonably well understood. Although there is much fine detail the main outlines are rather simple. Our understanding is based on three main types of observation, namely of the distances, numbers and velocities of the stars, which reveal both the shape and the overall rotation of the Milky Way.

The determination of distance in astronomy is a complex subject. Here we shall describe just two methods, one of which is the most precise for nearby stars while the other is used to establish the extragalactic distance scale. The first is the well known parallax method. It is based on the fact that a nearby star viewed from opposite sides of the Earth's orbit round the Sun would appear to

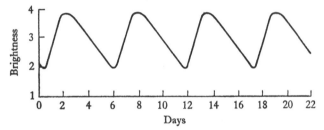

Fig. 5. The variation in brightness of the star Delta Cephei.

lie in different directions relative to background stars, by an amount depending on the distance of the star. The observed difference in direction then gives the distance of the star. This method has led to the parsec scale of distance. A star one parsec away has a parallax of one second of arc relative to the earth's orbit. Simple geometry then shows that a parsec is about 3 light-years or 3×10^{18} cm. Clearly to measure the parallax of a star 100 parsecs away requires the measurement of an angle of one-hundredth of a second of arc. This represents the limit of accurate measurement, so unfortunately the parallax method is useful only for relatively nearby stars.

The second method uses a type of star whose apparent luminosity varies periodically with time – the so-called Cepheid variables. The prototype is the star Delta Cephei whose light curve is shown in fig. 5. In 1912 Miss Henrietta Leavitt discovered that there is a well-defined relation between the period of a Cepheid variable and its absolute luminosity (as determined from measurements of distance for the nearest Cepheids and the inverse square law of apparent luminosity). In this period–luminosity relation the brightest stars have the longest periods (fig. 6). If there is reason to believe that a given variable belongs to the class for which this

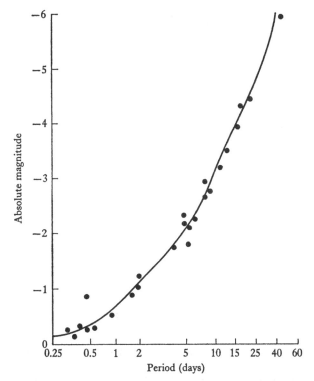

Fig. 6. The period–luminosity relation for Cepheid variables. The larger the absolute luminosity of a Cepheid (here measured logarithmically on the magnitude scale), the longer its period. A measurement of its period thus enables its absolute luminosity to be deduced. Its distance can then be inferred from its apparent luminosity.

relation has been established, its observed period determines its absolute luminosity and then its apparent luminosity enables its distance to be determined. This method has given trouble in the past because there are actually different period–luminosity relations for different classes of variable, but these difficulties have by now been mostly overcome. The method works out to distances at which the variables can still be adequately distinguished, that is, out to megaparsecs (one megaparsec is 10^6 parsecs). Such distances, as we shall see, take us well outside the Milky Way.

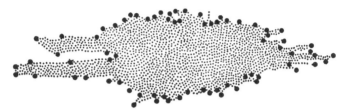

Fig. 7. Sir William Herschel's model of the Milky Way.

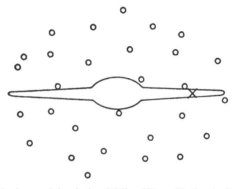

Fig. 8. Shapley's model of the Milky Way. Each circle represents a globular cluster. The hub and disc contain most of the stars in the Milky Way. The Sun, which is two-thirds of the way out from the centre, is marked by a cross.

Star counts were begun in the 18th century by Sir William Herschel who proposed a disc-like model for the Milky Way (fig. 7). A more modern version of this model is shown in fig. 8, which is due to Shapley (1918). Shapley's main innovation was to place the Sun near the edge of the system instead of at the centre where it had always previously been placed. Shapley discovered the eccentric position of the Sun while studying globular clusters – these are symmetrical star clusters which contain from 10 000 to

1 000 000 stars. The Cepheid variable method was used to determine the distances of about 100 globular clusters. From these distances, Shapley derived their distribution in space. They are arranged in a nearly spherical system whose centre is in the direction of Sagittarius (plate 3) at a distance from the sun which is now thought to be about 10 kiloparsecs. Most of the stars, however, are organised in a flattened system, reminiscent of Herschel's disc.

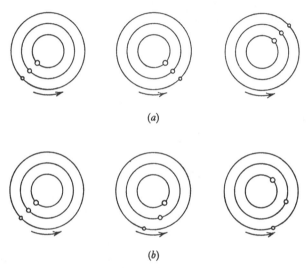

(a)

(b)

Fig. 9. (a) Rigid rotation, as in a wheel. Three points originally on a straight line remain on a straight line.

(b) Differential rotation, as in the solar system. Points near the centre move ahead of those farther out.

The real understanding of this flattened system began in 1926–7 when Lindblad and Oort discovered from the motions of the stars that the Milky Way is rotating. This rotation had previously been postulated by the philosopher Immanuel Kant in order to explain the flattened form of the Milky Way. Furthermore, Oort found, as Kant had suggested, that the Milky Way does not rotate like a rigid wheel but rather in the same way as the planets rotate about the Sun (fig. 9). The difference between these two types of rotation lies in the fact that, whereas in a rigid wheel every particle rotates with the same period, in the solar system the nearer a planet is to the Sun the shorter is its period. This is a characteristic property of circular motion around a central attracting mass. In fact much of the mass of the Galaxy is concentrated into its central regions.

It is fortunate that the stars move in this way since, if they all had the same period, their rotation would have been very difficult to detect. They would have had no radial motions relative to the Sun and a common transverse motion. This transverse motion can now be marginally detected with respect to external galaxies, but in the mid-twenties it would not have been possible to do this. As it is, the differential rotation of the stars leads to observable radial motions with a characteristic pattern shown in fig. 10. This pattern was discerned by Oort in 1927. From its orientation, amplitude and phase he was able to deduce that the axis of rotation is indeed at right angles to the disc, that the axis cuts the disc at a point lying in the direction of Sagittarius, in agreement

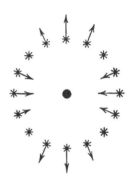

Fig. 10. The radial motions of nearby stars relative to the Sun.

with Shapley's results, and that the Sun makes a complete revolution in about 10^8 years. In conjunction with the distance to the centre, this enables the mass of the Galaxy to be estimated, just as we determined the mass of the Sun from the Earth's orbital period and the distance of the Sun. The mass of the Galaxy comes out to be of the order of 10^{44} grammes, corresponding to about 10^{11} solar masses.

It would not be appropriate to enter here into a more detailed account of the distribution and motion of the stars, except to mention one particular feature. As we shall see, many external galaxies have a characteristically spiral appearance (plate 5), while others are more or less featureless. It was established in 1952 that our Galaxy too is a spiral. Many bright young stars lie along spiral arms, in one of which the Sun itself resides. The origin of these arms is not completely understood but it is quite possible that what characterises them is actually an increased gas and dust density, the concentration of stars being a consequence of the resulting enhanced rate of star formation. This question of the gas and dust content of the Galaxy now claims our attention.

GAS AND DUST IN THE MILKY WAY

About 10 per cent of the mass of the Galaxy is in the form of interstellar gas and dust. These constituents were first discovered from the resulting scattering and absorption of starlight, but the best way to observe the gas nowadays is from its radio emission at a wavelength of 21 cm. The importance of this radio spectral line (which is emitted by atomic hydrogen) was first realised in 1944 by van de Hulst. He calculated that a sensitive radio receiver should be able to detect the 21 cm line as emitted by clouds of hydrogen gas in the Galaxy. This prediction was not verified until 1951, but by now our Galaxy (and many of its neighbours) has been thoroughly mapped at 21 cm. As fig. 11 shows, the main result is that the bulk of the atomic hydrogen in our Galaxy lies along spiral arms. A further basic result is that the rotation curve of the Galaxy (fig. 12) can be determined from the Doppler shift of the 21 cm line which tells us the radial velocity of each hydrogen cloud. This hydrogen does not, however, partake of circular motion near the centre of the Galaxy. For instance, an arm-like structure 3 kiloparsecs from the centre can be seen approaching us at a speed of 50 km s^{-1}. This and other radio evidence suggests that violent events are occurring or have occurred relatively recently at the galactic centre. Investigation of this intriguing problem is severely handicapped by the fact that the central regions cannot be observed optically at all, owing to the heavy obscuration in that direction.

In recent years other interstellar radio lines have been detected, from OH, ammonia, water vapour, formaldehyde and other molecules and from the recombination of ionised hydrogen, helium and possibly carbon. Some of these lines show remarkable physical behaviour, suggestive in many cases of maser-like action. The study of these phenomena is still in its infancy.

COSMIC RAYS

Cosmic rays are an important constituent of the Milky Way, not only because they serve as probes for the structure and properties of the interstellar regions of our Galaxy where they propagate but

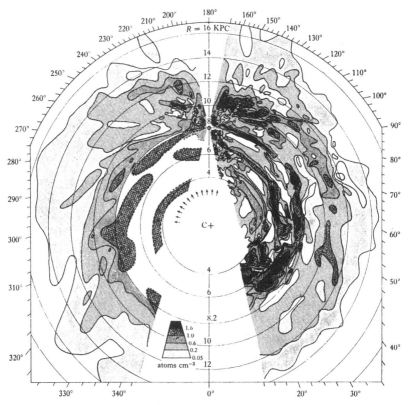

Fig. 11. The distribution of atomic hydrogen in the disc of the Galaxy, as derived from radio observations of the 21 cm line. (After J. H. Oort, F. J. Kerr and G. Westerhout, *Monthly Notices of the Royal Astronomical Society* **118**, plate 6 (1958).)

also *dynamically* in the sense that they exert a significant pressure on the interstellar gas. The cosmic rays impinging on the Earth were first detected in 1912, but their astrophysical significance was appreciated only recently. Most of them consist of high energy protons although, as we shall see, other types of particle are also represented. Most of the protons are *relativistic*, that is, their velocity v is close to the velocity of light c. Under these circumstances their mass m is given by Einstein's relation

$$m = \frac{m_0}{\sqrt{(1 - v^2/c^2)}},$$

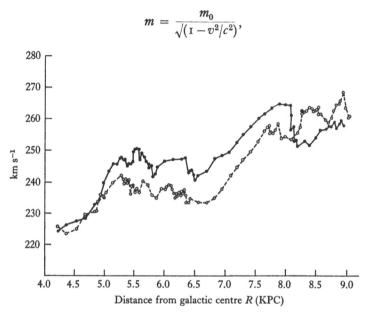

Fig. 12. Galactic rotation curves for northern (●——●) and southern (○ – – – – ○) sides of the Sun–centre line, as derived from 21 cm observations. (From F. J. Kerr, *Annual Review of Astronomy and Astrophysics* **7**, 39 (1969).)

where m_0 is their mass when they are at rest. Similarly, their total energy E is given by

$$E = \frac{m_0 c^2}{\sqrt{(1 - v^2/c^2)}}.$$

When v is very close to c the mass m is very much greater than the rest-mass m_0, and the energy E is very much greater than the rest-energy $m_0 c^2$. For a proton this rest-energy is about a billion electron volts (1 BeV $\sim 10^{-3}$ ergs), and it so happens that the mean energy of the cosmic ray protons is about 2 BeV, a fairly relativistic energy. However, individual cosmic ray particle energies go up to the fantastic value of 10^{11} BeV (10^8 ergs). By contrast the biggest accelerator working today (at Serpukhov) produces particles of 70 BeV, while the next generation of accelerators should increase this limit to 300 BeV. These accelerators have the advantage of producing much larger particle fluxes but there seems to be no prospect of their competing with nature to produce the highest energies.

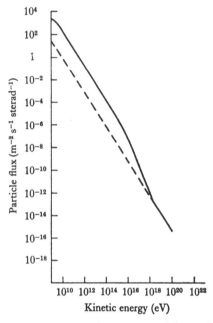

Fig. 13. The energy spectrum of cosmic rays. The dashed line represents a possible extragalactic component.

The main properties of the cosmic rays reaching the earth are as follows:

(i) Their flux is constant in time except at energies below 10 BeV where the influence of the solar system is important.

(ii) Their flux is the same in all directions, that is, they are isotropic.

(iii) The number of particles whose energy exceeds E BeV (the integral energy spectrum) is given by $\sim 5000\, E^{-1.5}\, \mathrm{m}^{-2}\, \mathrm{s}^{-1}\, \mathrm{sterad}^{-1}$ with fair approximation over the whole energy range from 10 BeV to 10^{11} BeV (fig. 13).

(iv) They comprise in addition to protons, α-particles and heavier nuclei up to atomic number 26 (iron). In addition some evidence has recently been found for much heavier particles up to atomic numbers exceeding 93 (transuranic elements).

These properties have the following significance:

(i) The cosmic ray flux almost certainly fills the Milky Way, and corresponds to an energy density in interstellar space of about 1 eV cm^{-3} (10^{-12} erg cm^{-3}). This is comparable with the energy density of starlight, the turbulent kinetic energy density of the interstellar gas and, as we shall see later, the energy density of the interstellar magnetic field. This is the basis of our statement that the cosmic rays are dynamically important. They constitute a relativistic gas whose energy and pressure cannot be ignored. The near-equality of the various energy densities is probably no accident but despite many attempts a full understanding of it has not yet been achieved.

(ii) The isotropy suggests that the cosmic rays do not travel in straight lines from their sources. This has led to the proposal that the interstellar gas is permeated by a magnetic field that deviates the paths of the charged cosmic rays from their original directions of motion. This proposal is now known to be correct as we shall see.

(iii) The energy spectrum tells us something about the mechanism that accelerates the cosmic rays in the first place, but this mechanism is not yet understood. Nor is the seat of the mechanism known, although it has been plausibly argued that it occurs mainly in supernova explosions and is perhaps related to the pulsars. We know from the laboratory, from processes occurring in interplanetary space, and from solar flares, that instabilities involving ionised gas (plasma) and magnetic fields lead to the acceleration of individual protons to high energies. Such processes may well occur in more violent fashion in supernova explosions, although we do not even know whether magnetic fields play an important part in such cases. However, if pulsars are responsible the acceleration mechanism may be quite different from that occurring in solar flares.

(iv) The chemical composition of the cosmic rays also tells us something about the acceleration process, but this information has not yet been used very effectively. What *has* been useful is the inference that the cosmic rays have passed through a known amount of material in reaching us from their sources. This is deduced from the abundances of lithium, beryllium and boron in cosmic rays. These abundances relative to hydrogen are millions of times greater than those measured in any celestial body, and it is supposed that these light elements occur in the cosmic rays because some of the heavier cosmic ray particles collide with the inter-stellar gas (or gas in the sources) and are broken up into lighter particles. In this way it is deduced that the cosmic rays have, on the average, passed through about 3 g cm^{-2} of hydrogen (that is, their path-length multiplied by the ambient density of hydrogen is 3 g cm^{-2}). In the disc of the Galaxy the average density of inter-stellar hydrogen is about 10^{-24} g cm^{-3}, so the corresponding path-length cannot exceed 3×10^{24} cm (and may be less since the sources may also contribute appreciably). Since the cosmic rays move at virtually the speed of light (3×10^{10} cm s^{-1}), they cannot have spent more than 10^{14} seconds or 3 million years in the disc. Pre-sumably in that time they are able to leak out of the disc despite the confining nature of the galactic magnetic field.

It is a matter of definition whether we include in the cosmic rays other types of high energy particles and radiation such as electrons, X-rays, γ-rays and neutrinos. We now consider these in turn.

RELATIVISTIC ELECTRONS

These were first detected in balloon flights above the atmosphere by Earl in 1961. They have a flux which is a few per cent of the cosmic ray proton flux. The rest-energy of an electron is much less than that of a proton, namely half a million electron volts ($\frac{1}{2}$ MeV) instead of 1 BeV, so electrons become relativistic at a corres-pondingly lower energy than do protons. The differential energy spectrum of the cosmic ray electrons (that is, the flux of particles per unit energy interval) is shown in fig. 14. Their importance stems from the fact that when they are relativistic and are moving in a magnetic field they radiate an appreciable flux of electro-magnetic waves that can be picked up by radio telescopes. This type of radiation is known as synchrotron radiation for the rather inadequate reason that electrons in a synchrotron type of accelerator move in a magnetic field and radiate electromagnetic waves. The radiated spectrum peaks at a frequency ν_m given by

$$\nu_m \sim \frac{1}{10} \left(\frac{E}{m_0 c^2}\right)^2 \frac{eB}{m_0 c},$$

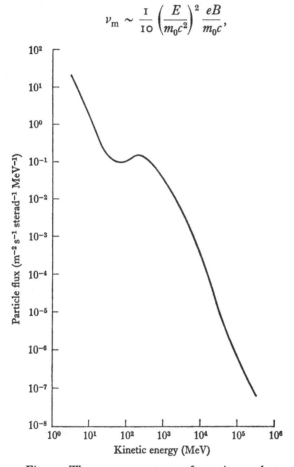

Fig. 14. The energy spectrum of cosmic ray electrons.

where E is the energy of the relativistic electron and B is the magnetic field strength. The full spectrum of the synchrotron radiation emitted by an electron is shown in fig. 15.

In an actual synchrotron $E/m_0 c^2$ might be of order 1000 and so with a magnetic field of 10^3 gauss the radiated power would be a maximum at a frequency $\sim 10^{15}$ Hz, that is, at an optical frequency. By contrast if a cosmic ray electron of energy 1 BeV moved in a magnetic field of 10^{-5} gauss, the peak frequency would be about 50 MHz, that is, a radio frequency. Now the Milky Way is known to emit radio noise whose frequency spectrum (fig. 16) and intensity are consistent with the observed electron energy spectrum and flux if the emission is due to the synchrotron process in a magnetic field of 5×10^{-6} gauss. Moreover this background is

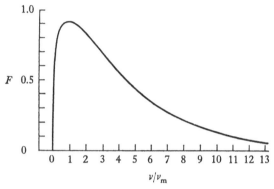

Fig. 15. The frequency spectrum of synchrotron radiation emitted by a single relativistic electron moving transversely to a magnetic field B. F must be multiplied by $2 \times 10^{-22} B_{\text{gauss}}$ to obtain the radiated power per unit frequency in c.g.s. units.

partially linearly polarised which would be expected on the synchrotron mechanism. In addition the existence of an interstellar magnetic field has now been definitely established by other methods. The synchrotron hypothesis for the radio emission from the Milky Way is thus almost certainly correct. If we may reverse the argument we can regard the radio emission as providing evidence that cosmic ray electrons fill the Galaxy with much the same flux as is observed at the Earth. This then strengthens the hypothesis that the cosmic ray protons also fill the Galaxy with the same flux as is observed at the Earth. We cannot test this hypothesis directly because owing to the larger mass of the proton, proton synchrotron radiation is too weak to observe.

The same analogy between protons and electrons suggests that the cosmic rays may actually fill a larger region than the galactic disc. The reason is that there is some evidence that radio emission comes not only from the disc of the Galaxy but also from a nearly spherical halo around the Galaxy of dimensions about 30 kiloparsecs. This halo is thought to contain hot ionised hydrogen of density ∼ 10^{-27} g cm^{-3} and temperature ∼ 10^6 °K, although this hydrogen has not been directly detected. If the galactic magnetic

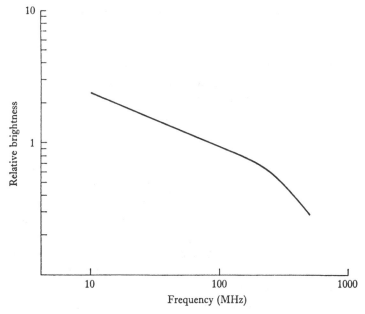

Fig. 16. The radio spectrum of the Galaxy.

field links on to a field in the halo, then cosmic ray electrons in the disc of the Galaxy could be guided into the halo where they would continue to emit synchrotron radiation. Presumably cosmic ray protons and heavier nuclei would follow a similar path into the halo. At the time of writing the *radio* evidence for a halo is not well established, but it is quite likely that a *physical* halo exists, that is, a region of ionised gas, magnetic fields, and cosmic rays. If the field strength and the electron flux are weaker than in the galactic disc the radio emission from the halo would be correspondingly harder to detect. This is natural enough, but we badly need positive evidence that a significant halo exists.

Another likely source of electron synchrotron radiation is the Crab nebula, since it is a radio source which is linearly polarised. The intensity of the source is such that the electron flux density within it must be much greater than in the Milky Way as a whole. This fits in with the general idea that cosmic rays are accelerated in supernovae, and this idea can be made plausible quantitatively, as has been shown especially by Ginzburg and Syrovatsky. The synchrotron hypothesis for the Crab was beautifully confirmed following Shklovsky's suggestion that its optical emission may also be synchrotron in origin but produced, of course, by much higher energy electrons than the radio noise. This hypothesis would require the optical emission also to be linearly polarised, as was then found to be the case by Oort and Walraven, and by Baade (plate 6).

Our understanding of the Crab nebula has recently been greatly increased by the discovery that the pulsar in the Crab (p. 11) is slowing down at the rate of about 1.35×10^{-5} seconds per year. If the pulsar is a neutron star of about one solar mass and radius 10 kilometres rotating with a period of 33 milliseconds then this rate of slowing down would imply that the neutron star is losing rotational energy at a rate of about 10^{38} erg s^{-1}. This result is of great significance because it has long been known that the Crab requires a power input of this order *at the present time* in order to account for its observed synchrotron radiation. In particular the very high energy electrons responsible for the optical synchrotron radiation emit energy so rapidly that they have an effective lifetime far shorter than the age of the Crab. There is no possibility of these electrons having been accelerated at the time of the supernova outburst in 1054. It is therefore a very attractive hypothesis to suppose that the pulsar is the powerhouse of the Crab. We may extend this hypothesis by supposing that most of the cosmic rays in the Galaxy are accelerated in pulsars associated with other supernova remnants. However, it remains possible that most of the cosmic rays are accelerated during the supernova explosion itself.

X- AND γ-RAYS

The distinction between X- and γ-rays is a somewhat arbitrary one, but for convenience we may take the dividing line to be an energy of 100 keV (kiloelectron volts). The first celestial X-rays (other than solar X-rays) were detected in 1962 by R. Giacconi, H. Gursky, F. Paolini and B. Rossi. About 50 discrete X-ray sources are now well established, most of which lie fairly close to the plane of the Galaxy. Two of them have been identified with stars which appear to be old novae. In addition the Crab nebula is an X-ray source. The remainder probably also lie in our Galaxy, with the exception of the radio galaxy Virgo A (see chapter 4), the quasar 3C273 (chapter 5) and several Seyfert galaxies (chapter 4). Superposed on these discrete sources is a general background which may be of extragalactic origin. Its energy spectrum is shown in fig. 17. It is not known what produces this background, but there have been some interesting cosmological speculations that are discussed in chapters 10 and 15.

Attempts to find celestial γ-rays have been less successful. Only very recently (late 1968) has a positive measurement been made. It appears that a large flux of γ-rays of 100 MeV is coming from the centre of our Galaxy and from its disc. The origin of this flux is unknown.

NEUTRINOS

As we saw in chapter 1, attempts are being made to detect neutrinos from the Sun. More truly cosmical neutrinos may also exist, and have been much speculated upon, but there are no observations as yet. Their weak interaction with matter makes them exceedingly difficult to detect.

GRAVITATIONAL WAVES

It has recently become apparent that the Galaxy may contain another type of weakly interacting radiation, namely gravitational waves. According to Einstein's general theory of relativity, when a particle is accelerated or a body is deformed in shape, gravitational waves are emitted. These waves travel through empty space with the speed of light, carrying with them the information that a change has occurred in the gravitational field of the source. This prediction of general relativity has not yet been conclusively

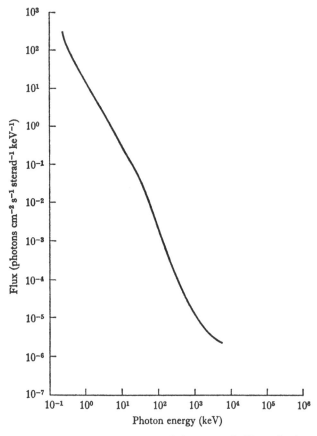

Fig. 17. The energy spectrum of the isotropic X-ray background.

verified by experiment, although one would expect that any relativistic theory of gravitation would lead to a similar prediction. The difficulty facing the experimenter arises from the weakness of the gravitational interaction. An example of this weakness is the gravitational force between an electron and a proton, which is only about 10^{-40} of their electrical force (cf. p. 124). It needs a very massive body, such as the Earth, to produce an easily measurable gravitational force. It follows from this that one can generate in the laboratory only very weak gravitational waves, which in turn lie below the threshold of detection by even the most modern techniques.

The question then arises whether naturally occurring gravitational radiation is intense enough to be detectable. Until recently the prospects would not have seemed promising. However, it is now realised that many types of objects in the Universe can explode violently, leading to the release of very large amounts of energy. We shall meet several examples of these violent outbursts in this book, but our immediate concern is that such outbursts may also lead to the release of a detectable flux of gravitational radiation.

Inspired by this challenge, Professor J. Weber of the University of Maryland, has devoted many years to the design of gravitational wave detectors of high sensitivity. A typical detector consists of a metal bar about a metre long suspended by a wire in a high vacuum. When a gravitational wave is incident on the bar it is set oscillating, and these oscillations are detected by piezoelectric crystals bonded to its surface. Great care must be taken to isolate the bar from non-gravitational disturbances. Some idea of the delicacy of Weber's equipment may be gained from the fact that he is able to measure an oscillation of the bar when its amplitude is only 10^{-14} cm. Since thermal noise alone gives rise to oscillations of this amplitude, the signal from a gravitational wave would have to correspond to larger amplitudes in order to be identified .This represents a serious limitation of the sensitivity of the detector. Weber has overcome this limitation by setting up several similar detectors, and looking for coincidences in their responses.

In mid-1969 Weber reported for the first time that he had obtained sufficient coincidences to be able to rule out thermal noise as the responsible agent. Moreover he found as many coincidences (within a resolution time of half a second) between detectors a thousand kilometres apart as between detectors in the same laboratory. The coincidence rate was about one per week, and he felt able to state that many of the coincidences could not be due to any agency other than gravitational waves.

S M C

The implied flux of gravitational radiation is considerable. Moreover, Weber's recent observations indicate that this flux is greatest when the centre of the Galaxy is in the beam of his detectors. This would suggest that catastrophic processes occurring in the galactic centre are responsible for his events, but the rate of about one per week (or one per day in his later experiments) seems very high. Indeed the mass lost by the Galaxy in gravitational radiation would become an important item in its overall mass balance and dynamics, unless the events are of relatively recent origin. The discovery of gravitational radiation is of such fundamental importance to physics, and the flux reported by Weber is so large, that it is essential to confirm his measurements by independent experiments. Several such experiments are now under way, and should soon be giving results.

INTERSTELLAR MAGNETIC FIELDS

We have already mentioned interstellar magnetic fields several times, but they pervade so much of modern astrophysics, and they play such an important role in later parts of this book, that they deserve a separate heading of their own. Leaving aside for a moment the question of the origin of the magnetic field, let us list the reasons why a magnetic field is believed to permeate the interstellar gas.

(*a*) It would account for the isotropy of the cosmic rays.

(*b*) After many attempts the (Zeeman) influence of the magnetic field on the 21 cm line of atomic hydrogen has recently been detected.

(*c*) The light from some stars is linearly polarised. This is best explained in terms of scattering by elongated magnetised dust particles that are alined by the interstellar magnetic field.

(*d*) The radio emission from the Milky Way is probably due to synchrotron radiation, which requires an interstellar magnetic field.

(*e*) Extragalactic radio sources (chapters 4 and 5) are often linearly polarised and the plane of polarisation is found to vary with the wavelength of observation. The manner of this variation is strongly suggestive of Faraday rotation of the plane of polarisation by a magnetic field. Such a mechanism would require there to be free electrons in a magnetic field somewhere along the line of sight. The amount of Faraday rotation is found to be closely correlated with galactic latitude, that is, with the length of the line of sight that lies in the disc of our Galaxy (fig. 18). This correlation implies that there must be a magnetic field in the Galaxy, and is the most convincing evidence that such a field exists. Indeed a beginning is now being made in the task of mapping out the detailed

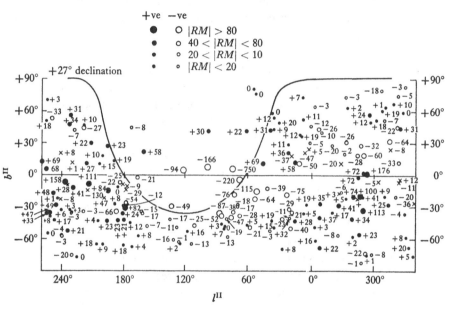

Fig. 18. The distribution of the Faraday rotation-measures (RM) of extragalactic radio sources with galactic longitude l^{II} and latitude b^{II}. The sources north of declination $+27°$ are not detectable from Parkes, and their rotation measures are taken from other surveys. (After Gardner *et al.*, *Austr. J. Phys.* **22**, 813 (1969).)

configuration of the magnetic field. This task may be much simplified by the recent discovery of Faraday rotation effects in pulsars.

The intensity of the interstellar magnetic field can be estimated from (*b*), (*c*), (*d*) and (*e*) and the value that is found is about 5×10^{-6} gauss. This is very low by terrestrial standards but on the cosmical scale can be of decisive importance. The energy density of such a field is about 1 eV cm^{-3}, a value we have already quoted in connection with the energy density of the cosmic rays. An individual cosmic ray of energy E electron volts moves around the magnetic field in a spiral whose (Larmor) radius in centimetres is given by $E/300B$, where the magnetic field B is in gauss. The galactic magnetic field thus has an important effect on the motion of the cosmic rays if their Larmor radius is substantially less than the size of the Galaxy. This condition breaks down when E exceeds about 10^{17} electron volts, and such very high energy cosmic rays may have an extragalactic origin and even fill the whole Universe with the same flux as is observed at the Earth.

The problem of the origin of the galactic magnetic field is unsolved. One thing that is understood is that the currents giving rise to the magnetic field meet very little electrical resistance in the interstellar gas, and so take a long time to decay. A better way of expressing this is to say that the self-inductance of the interstellar medium is very large. This means that once a magnetic field were formed it would take a very long time to decay, longer in fact than the age of the Galaxy. By the same token, however, it is difficult to form the field in the lifetime of the Galaxy. It has been suggested that a small seed field can be amplified by turbulent fluid motions in the interstellar gas, but the theory of such a process is very difficult to work out and it remains little more than a suggestion. Moreover (*c*), (*d*) and (*e*) show that the direction of the magnetic field has large-scale order in the Galaxy, which is hard to reconcile with a turbulent origin.

An alternative suggestion, which is of great interest from the cosmological point of view, is that the magnetic field was already present when the Galaxy was formed. Of course this only transfers the problem, but it may be that conditions in the Universe in its early stages were sufficiently different from those obtaining today as to facilitate the formation of magnetic fields. On this speculative note we close this chapter, preserving the thought that what is by now a familiar and well-explored astrophysical property of our own Galaxy may have a completely cosmological explanation.

3 External galaxies and the expansion of the Universe

We must now face the fundamental question: does the Milky Way comprise the total contents of the Universe? There are theoretical reasons, connected with what is known as Mach's principle (chapter 8), for believing that there must be an enormous amount of matter outside the Milky Way, but the first suggestion that such matter actually exists appears to have been made on more empirical grounds by Kant, whose comments on the flattened shape of the Milky Way we have already met. Kant suggested that the objects then known to astronomers as nebulae because of their cloudy unresolved appearance were actually galaxies lying outside the Milky Way system and having a somewhat similar structure. Since Kant's time it has been discovered that these nebulae are of several kinds: in particular astronomers have distinguished between those that can be resolved into stars by powerful telescopes and those that are simply luminous gas. Of the resolved nebulae some are globular clusters of stars while others have a flattened and spiral appearance.

A great controversy arose as to whether the spiral nebulae were inside or outside the Milky Way. In view of the similar controversy that is now raging about the whereabouts of the so-called quasi-stellar objects (chapter 5), it is interesting to recall some of the opinions expressed in that old controversy. In 1905 Agnes Clerke, a historian of astronomy, wrote as follows:

The question whether nebulae are external galaxies hardly any longer needs discussion. It has been answered by the progress of research. No competent thinker, with the whole of the available evidence before him, can now, it is safe to say, maintain any single nebula to be a star system of co-ordinate rank with the Milky Way.

The main evidence for this view was the fact that the distribution of nebulae is closely related to the structure of the Milky Way – all the nebulae lie in directions well away from its disc. We now know, however, that this feature of their distribution is only apparent. It is caused by the dust that lies in the disc, an explanation already proposed by the English astronomer Sir Arthur Eddington in 1914, sixteen years before it was established that the dust actually exists:

In the days before the spectroscope had enabled us to discriminate between different kinds of nebulae, when all classes were looked upon as unresolved star clusters, the opinion was once widely held that these nebulae were 'island universes', separated from our own stellar system by a vast empty space. It is now known that the irregular gaseous nebulae, such as that of the Orion, are intimately related with the stars, and belong to our own system; but the hypothesis has recently been revived so far as regards the spiral nebulae....

It must be admitted that direct evidence is entirely lacking as to whether these bodies are within or without the stellar system. Their distribution, so different from that of all other objects, may be considered to show that they have no unity with the rest. Indeed, the mere fact that spiral nebulae shun the galaxy may indicate that they are influenced by it. The alternative view is that lying altogether outside our system, those that happen to be in low galactic latitudes are blotted out by great tracts of absorbing matter similar to those which form the dark spaces of the Milky Way.

If, however, it is assumed that these nebulae are external to the stellar system, that they are in fact co-equal with our own, we have at least an hypothesis which can be followed up, and may throw some light on the problems that have been before us. For this reason the 'island universe' theory is much to be preferred as a working hypothesis; and its consequences are so helpful as to suggest a distinct probability of its truth.

The argument was continued between Harlow Shapley and H. D. Curtis in the years 1917 to 1921. In this latter year Shapley wrote:

It seems to me that the evidence, other than the admittedly critical tests depending on the size of the galaxy, is opposed to the view that the spirals are galaxies of stars comparable with our own. In fact, there appears as yet no reason for modifying the tentative hypothesis that the spirals are not composed of typical stars at all, but are truly nebulous objects.

In the same year Curtis wrote:

I hold, therefore, to the belief that the galaxy is probably not more than 30 000 light years in diameter; that the spirals are not intra-galactic objects but island universes, like our own galaxy, and that the spirals, as external galaxies, indicate to us a greater universe into which we may penetrate to distances of ten million to a hundred million light-years.

THE HUBBLE ERA (1924–36)

The argument was resolved by Edwin P. Hubble in 1924. Using the 100-inch telescope at Mount Wilson, Hubble found that there were Cepheid variables in the great spiral nebula in Andromeda (plate 4) and also in other spirals. He was then able to determine their distance by using the period–luminosity relation. In this way he obtained a distance of 800 000 light-years for the Andromeda nebula, and similar values for other spiral nebulae. Now that these nebulae are established as stellar systems outside our own, we shall henceforth call them galaxies. In the years that followed Hubble studied these galaxies in great detail and in 1936 he summarised this work in his book *The Realm of the Nebulae*.

One of Hubble's main achievements was to establish methods of measuring the distances to galaxies which are too far away for their Cepheid variables to be detected. He did this in a series of steps, starting with the measurement of distances by means of Cepheid variables. This first step works only for the nearest galaxies, which are clustered together into what is known as the 'local group', to which the Milky Way belongs.

For his next step Hubble used supergiant stars, which are intrinsically brighter than Cepheids. These stars can be detected in many distant galaxies whose Cepheids are invisible. Hubble assumed that the brightest supergiants in all the galaxies have about the same absolute luminosity; their apparent luminosity then indicates their distance. Hubble checked this method by using it on the local group, whose distances were already known from their Cepheid variables, and then applied it to more distant galaxies. In this way he extended the range of measured distances from 1 million light-years to 10 million light-years.

This was as far as Hubble could go using stars as distance indicators. All that was left for him in the final step was to turn to the galaxies themselves, and to assume that they all have the same absolute luminosity. That this step is a reasonable one was indicated by his observations of a large cluster of galaxies in the direction of the Virgo constellation – a cluster that he estimated to be about 8 million light-years away. His measurements of the apparent luminosity of galaxies in this cluster showed that their absolute luminosities differ from one another by at most a factor of 10. The assumption that all galaxies have an absolute luminosity in the middle of this range of values would then lead to an error in their distance of about a factor 3 in unfavourable cases. The total range of distances is fortunately so large that this uncertainty would not conceal any systematic trend in the properties of galaxies at different distances. We now know that galaxies have a much wider range of absolute luminosities, but at least when dealing with clusters of galaxies one can reduce the uncertainty by restricting oneself to the brightest galaxies in the cluster.

By using these methods, Hubble explored the Universe out to the great distance of 500 million light-years – a region that contains about 100 million galaxies. This is a considerably greater number than the 500000 that were known before the days of the 100-inch telescope. Hubble's book contains many details about some of these galaxies, but we can describe only his main results (with the warning that his distance-scale has been considerably modified in recent years). According to Hubble, the average distance between galaxies is about a million light-years. He also concluded that they are, on the whole, much smaller than the Milky Way, whose diameter he gave as 100000 light-years. By comparison the majority of galaxies were considered to be only about 10000 light-years across, which would mean that they are separated by a distance 100 times their own size.

This did not mean, however, that the galaxies were more or less uniformly distributed with the spacing of a million light-years. On the contrary they show considerable clustering, ranging from pairs of galaxies through clusters with fifteen or twenty members like the local group, up to clusters such as the one in Virgo containing several thousand galaxies. It is not yet known whether these clusters are gravitationally bound systems. In many cases the galaxies themselves are not sufficiently massive to bind the clusters, but there may be enough faint stars or gas between the galaxies to provide the necessary 'missing matter'. This question is also important in relation to the overall mean density of matter in the Universe, a quantity that plays a key role in discussions of the evolution of the Universe (chapter 8). According to recent estimates the mean density contributed by known galaxies lies between 10^{-31} and 10^{-30} g cm^{-3}. As we shall see, it is of critical importance to know how much this estimate should be increased because of the existence of undetected matter in galaxy clusters and also in the space between clusters. This key problem is still unsolved (chapters 9 and 10).

In addition to studying the distribution of galaxies in space, Hubble classified them into various types, according to their general appearance. The main types are spiral, barred spiral, and elliptical, with a few per cent of irregular shape (plate 7). The problem of understanding these different forms – whether they are related to age or some other property of a galaxy, and why the galaxies are distributed among these types in the observed proportions – is still unsolved.

Since 1936 many detailed studies of individual galaxies have been made, but on the whole the picture presented by Hubble still stands. The one major modification concerns his use of the Cepheid variable method of determining distances. It will be recalled that this method is based on the fact that there is a relation between the period of a Cepheid variable and its absolute luminosity – the period–luminosity relation. Now Walter Baade, an American astronomer, discovered in 1952 that there are two types of Cepheids with different period–luminosity relations (fig. 19). Distances within the Milky Way are unaffected by this discovery – they had been determined from Type II Cepheids, and so are in fact correct. Unfortunately, the Cepheids observed in other galaxies are of Type I, which means that their absolute luminosity had been underestimated. In consequence their distances had also been underestimated.

The change has been a substantial one. Distances derived by Hubble should be multiplied by about 5, thus greatly increasing the scale of the Universe. One consequence of this is that the Milky

Way is no longer a giant among galaxies but is of roughly average size for its luminosity, since the diameters of galaxies as deduced from their angular diameters must also be multiplied by a factor 5. In this way we have lost the last vestiges of what was once believed to have been our privileged status in the Universe. As Copernicus dethroned the Earth, and Shapley the Sun, so Baade dethroned the Milky Way. Since the local group of galaxies is a comparatively small cluster, the geocentric picture of the Universe is now completely discredited.

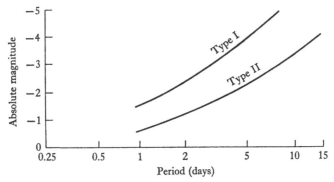

Fig. 19. The period–luminosity relation for the two types of Cepheid variables. The original relation (fig. 6) referred to Type II Cepheids only.

THE EXPANSION OF THE UNIVERSE

The first successful observation of the radial velocity of a galaxy from the Doppler effect of its spectral lines was made in 1912 by V. M. Slipher of the Lowell Observatory. He found that the Andromeda galaxy is approaching the earth at a velocity of about 200 km s^{-1}. When we recall that most stars move no faster than about 50 km s^{-1}, it will be seen that this is a remarkable result. Slipher went on to measure the spectra of other galaxies and found that most of them show a red shift, which means that, unlike Andromeda, they are receding rather than approaching. The amount of the shift again implied large velocities. By 1914, Slipher had measured the spectra of thirteen galaxies, all but two of which are receding at around 300 km s^{-1}.

These velocities were by far the largest that had ever been

measured in astronomy. But worse was yet to come. By 1917 velocities of 600 km s^{-1} had been registered, and even this record was soon surpassed. It is interesting to read a contemporary comment. Eddington wrote in 1923:

One of the most perplexing problems of cosmology is the great speed of the spiral nebulae. Their radial velocities average about 600 kilometres per second and there is a great preponderance of velocities of recession from the solar system. It is usually supposed that these are the most remote objects known (though this view is opposed by some authorities), so that here if anywhere we might look for effects due to general properties of the world.

Eddington then gave a list of the radial velocities of spiral galaxies as measured by Slipher up to February 1922 and continued:

The great preponderance of positive (receding) velocities is very striking; but the lack of observations of southern nebulae is unfortunate, and forbids a final conclusion. Even if these also show a preponderance of receding velocities the cosmogonical difficulty is not entirely removed....
It will be seen that two nebulae (including the great Andromeda nebula) are approaching with rather high velocity and these velocities happen to be exceptionally well determined.

Eddington's words remind us that at that time it had not yet been definitely established that the spiral galaxies lie outside the Milky Way. Hubble's discovery that they do dates from the next year, 1924. Further light on Slipher's velocities was shed by the discovery in 1926–7 that the Milky Way is in rotation. The Sun's velocity around the centre of the Milky Way is about 250 km s^{-1}. Other objects in the Milky Way are also moving around its centre, so that their radial velocity relative to the Sun is much less than 250 km s^{-1} (cf. fig. 10). But objects outside the Milky Way do not share its rotation, so that their measured velocities must be corrected for the Sun's motion if we want to know their velocity relative to the Milky Way as a whole (fig. 20). When this correction was made the rapidly approaching galaxies which so worried Eddington were much slowed down. The irony of this is that after the correction the Andromeda galaxy has a velocity of approach of only about 100 km s^{-1}. Thus the first velocity measured by Slipher, which at the time seemed startlingly large, is actually a poor guide to the surprises that were to come.

The significance of Slipher's results was further clarified by Hubble's important discovery that the velocities of recession are by no means random. By using his measurements of the distances to spiral galaxies, Hubble established in 1929 that out to 6 million light-years the velocity of a galaxy is proportional to its distance (fig. 21). At first sight it might appear that the privileged status of the Milky Way had been restored by Hubble's discovery. How-

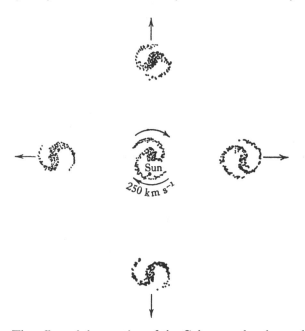

Fig. 20. The effect of the rotation of the Galaxy on the observed radial velocities of external galaxies.

ever, it was quickly realized that Hubble's result does not imply that the Milky Way is a unique centre of repulsion. On the contrary, a law of expansion in which velocity is just proportional to distance implies that any galaxy may be regarded as the centre of expansion, and would observe the same law of recession (fig. 22).

Hubble believed that the constant of proportionality in his law of recession was about 500 km s^{-1} Mpc^{-1}. This scale of velocities can be put in a more striking way, which serves to explain why

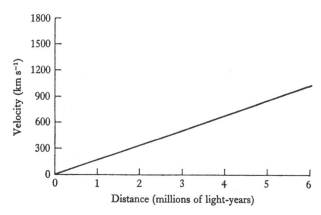

Fig. 21. Hubble's first velocity–distance relation for galaxies. The velocity of recession is just proportional to the distance of a galaxy. Hubble's distances should be multiplied by a factor of about 5.

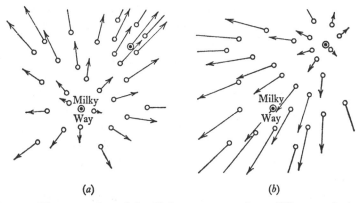

(a)   (b)

Fig. 22. The expansion of the Universe as seen from different galaxies, according to Hubble's law.

(a) The recession velocity of a galaxy is proportional to its distance from the Milky Way.

(b) The expansion of the Universe as seen from another galaxy. The recession velocity is still proportional to the distance. Thus Hubble's law does not favour the Milky Way.

Hubble's result created a great sensation at the time. Let us trace the motion of the galaxies backward in time, assuming that each galaxy has a constant velocity. Then Hubble's result implied that 2 billion years ago all the galaxies were crowded on top of one another. This was striking not only in itself but because the ages of the Earth and the Sun were believed to be greater than 2 billion years.

Of course the assumption that the Universe has been expanding at a constant rate may be false. In that case the crowding together may have occurred more than 2 billion years ago. This question could not be decided without a theory of the expansion. Meanwhile many people felt that the time of 2 billion years, which became known as Hubble's constant, had a basic significance for the Universe as a whole.

This reaction may seem to have been somewhat premature but later work tended to support it. By 1931, Hubble extended the range of validity of his law from 6 million to 150 million light-years. Finally ,with the help of Milton L. Humason's new measurements of Doppler shifts, Hubble reached out to 240 million light-years, where the velocities of recession are about one seventh the velocity of light. This was the situation when Hubble published his book, *The Realm of the Nebulae*, in 1936.

Since then the 200-inch telescope has come into operation on Mount Palomar and improved techniques have been devised for detecting the light that is collected by the telescope. This has enabled the red shifts of fainter, more distant, galaxies to be determined. However, the only important change in Hubble's results has come from the recognition of the large error in his distance scale. Hubble's constant is now believed to be about 10 billion years. This is greater than the accepted ages of the Earth and Sun, and comparable with the ages of the oldest star clusters. Thus there is no longer any difficulty in supposing that the Universe was once very dense.

A further recent development is that many radio galaxies (chapter 4) whose optical counterparts are often the brightest galaxy of a cluster have had their red shifts measured. These galaxies are so far away that the only distance indicator is their apparent luminosity. Allan Sandage has found that this method of estimating distance is reliable since the relation between red shift and apparent luminosity for radio galaxies is a well-defined one (fig. 23). This implies that they all have much the same absolute luminosity. Unfortunately, this is certainly not the case for quasi-

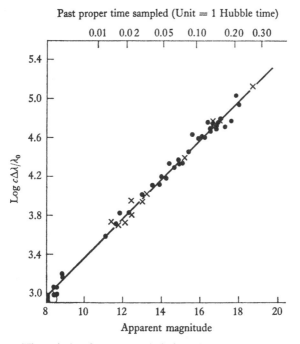

Fig. 23. The relation between red shift and apparent magnitude for the brightest galaxies in 42 clusters: ×, radio; ●, non-radio. Hubble's original (1929) linear relation of fig. 21 is represented in this diagram (which is due to A. R. Sandage, *Observatory* **88**, 99 (1968)) by the black rectangle in the bottom left-hand corner.

stellar objects (chapter 5), so that despite their very large red shifts they cannot be used to examine the Hubble law at greater distances.

Sandage has hopes of finding a deviation from the simple linear Hubble law at the largest red shifts that are available to him. Such a deviation would be of great theoretical interest, and would help to decide which is the best of the various cosmological models (chapter 8). Unfortunately the task is a difficult one because the distant galaxies are being observed a long time in the past, owing to the time it takes for their light to reach us. At such earlier times their absolute luminosity may not have been the same as it is now. Any systematic evolution in the intrinsic properties of the galaxies would contribute to the observed deviation from the simple Hubble law, and could easily mask the cosmological effect being looked for. Unfortunately we do not know enough about the evolution of galaxies to make the necessary corrections in a reliable way. This difficult problem is likely to remain unsolved for a long time to come.

4 Radio galaxies

INTRODUCTION

INTRODUCTION
From time to time in the development of science a discovery is made that is totally unexpected. Often such a discovery leads to the formulation of a new law of physics. In astronomy, however, it usually shows that we have failed to work out some of the consequences of the laws already known. The unexpected discovery of radio galaxies, and of the quasi-stellar objects described in the next chapter, are probably examples of such failures. As such they are dramatic reminders of the enormous gaps in our understanding of the behaviour of matter in bulk. Take, let us say, a mass of 10^{44} grammes – the mass of a typical galaxy – and ask how it will behave. Astronomers had been studying this question for a long time, but none of them dared to suggest that a substantial fraction of the mass may sometimes become involved in a violent explosion, liberating the total *rest-energy** residing in perhaps 10^{39} grammes of the material into the form of relativistic particles and magnetic fields, which then shines forth as a radio source of unprecedented power and size.

These, however, are the facts, and they are still not understood. We owe their discovery to the combined efforts of radio and optical astronomers over the last two decades in what is the first of the great revolutions in astronomy of our time. Other revolutions are now occurring, in X-ray astronomy, infra-red astronomy, and in the study of the microwave background (described in chapter 14), but credit for shaking the optical astronomer's faith that he had discovered essentially all the types of object that there are in the sky must go to the radio-astronomer. This revolution is still so

* See p. 23

recent that in describing it we shall adopt a historical approach. We do this not so much in order to allocate credit correctly to the people concerned (that, alas, is an extremely difficult task even in contemporary science) but to show the reader how in our own time a great branch of science can develop from small beginnings.

THE DISCOVERY OF RADIO SOURCES

The science of radio-astronomy began in an extremely small way before the Second World War when Karl Jansky succeeded in detecting a weak background of radio emission which we now know came from the Milky Way. However, estimates made at the time led people to believe that our Galaxy would be too weak a radio source to detect, and Jansky's results were not followed up. Fortunately when the war ended and physicists were released from their specialist activities a number of radar sets were also released and some of the physicists began to use them to survey the radio sky. Already in 1944 Grote Reber found indications of the existence of a discrete radio source in Cygnus, now called Cygnus A, in addition to the general background. However, the first clear evidence for the existence of such a discrete source was obtained in England in 1946 by J. S. Hey, S. J. Parsons and J. W. Phillips. This marks the beginning of the study of radio galaxies, although at the time this was not realised. The position of Cygnus A was far too uncertain for an identification to be possible with an optical object. Indeed, as we shall see, although Cygnus A is the second brightest object in the radio sky, it was not identified until 1954, another very significant date in this history.

The existence of Cygnus A was fully confirmed in 1948 by the work of M. Ryle and F. G. Smith at the Cavendish Laboratory and of J. G. Bolton and G. J. Stanley in Australia. In the same year Bolton discovered six other discrete sources including Taurus A and Centaurus A. With these successes in hand radio-astronomers began to conduct systematic surveys so as to be able to compile catalogues of radio sources. The first two catalogues were published in 1950, one by Stanley and O. B. Slee containing 18 sources, and a more extensive one by Ryle, Smith and B. Elsmore containing 50 sources. This latter catalogue is known as 1C,

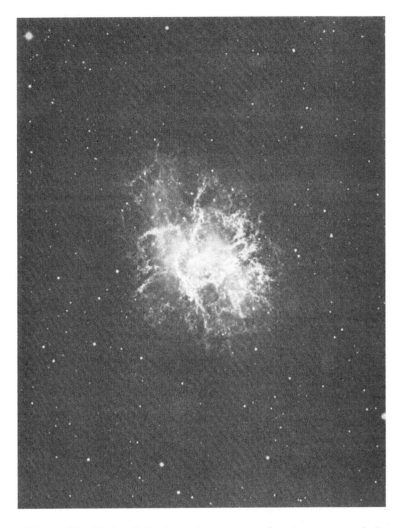

Plate 1. The Crab nebula. It is the remnant of a supernova explosion recorded by the Chinese in A.D. 1054. (Photograph from the Hale Observatories.)

[*Facing p.* 50]

Plate 2. The Milky Way, from Sagittarius to Cassiopeia. (Photograph from the Hale Observatories.)

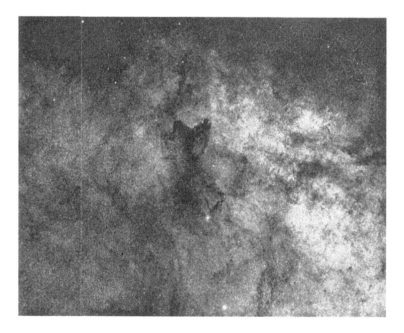

Plate 3. Clouds of stars in the region of Sagittarius. The centre of the Milky Way lies behind these clouds at a distance of about 10 kiloparsecs. (Photograph from the Hale Observatories.)

Plate 4. The Andromeda galaxy. It is a neighbour of the Milky Way and is probably similar to it in structure. (Photograph from the Hale Observatories.)

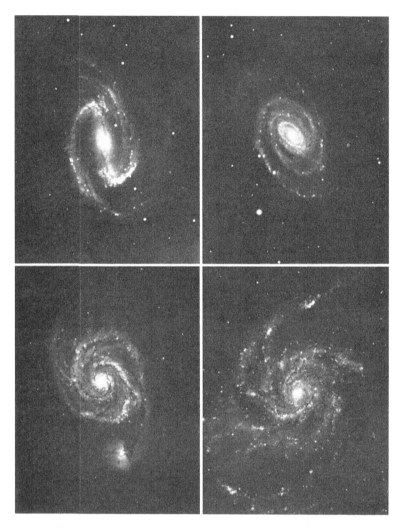

Plate 5. Four spiral galaxies. (Photographs from the Hale Observatories.)

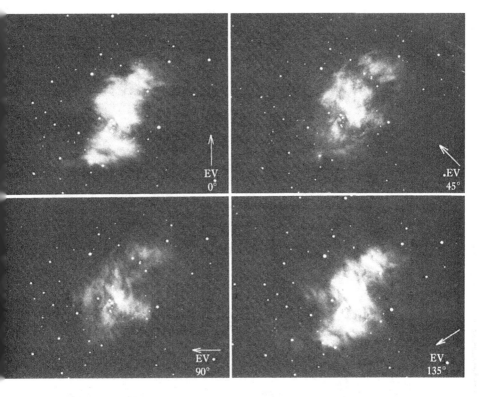

Plate 6. The Crab nebula observed in polarised light, with the electric vector at 0°, 45°, 90° and 135°. (Photographs from the Hale Observatories.)

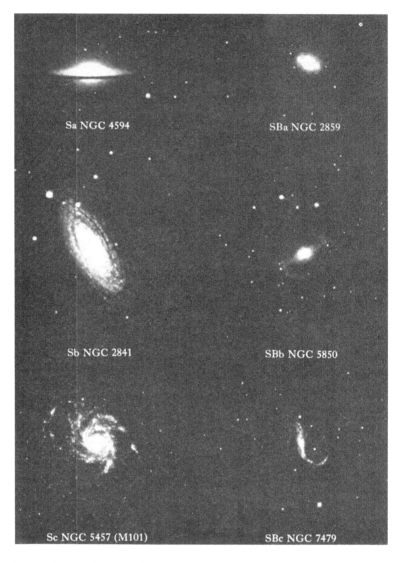

Sa NGC 4594

SBa NGC 2859

Sb NGC 2841

SBb NGC 5850

Sc NGC 5457 (M101)

SBc NGC 7479

Plate 7. The main types of galaxy, as classified by Hubble. There are spirals (Sa, b, c), barred spirals (SBa, b, c), ellipticals (Eo,..., 7) and a few per cent of irregular shape. The letters NGC and M refer to catalogues of celestial objects.

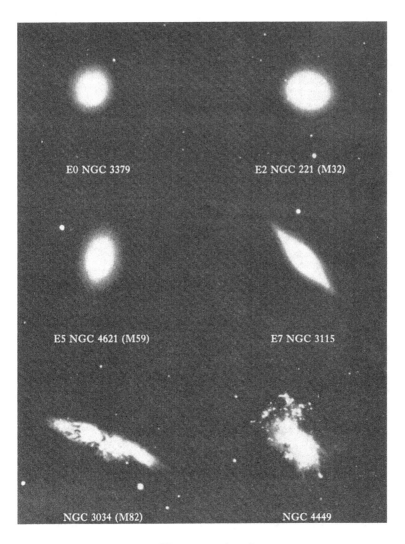

E0 NGC 3379

E2 NGC 221 (M32)

E5 NGC 4621 (M59)

E7 NGC 3115

NGC 3034 (M82)

NGC 4449

Plate 7, continued.

Plate 8. The Coma cluster of galaxies. (Photograph from the Hale Observatories.)

the first Cambridge catalogue. The positions of the sources had probable errors of several minutes of arc in right ascension and about a degree in declination.

The first really extensive survey was made in Cambridge in 1955. The resulting catalogue, known as 2 C, contained 1936 sources. This catalogue was the basis of the first attempt to draw cosmological conclusions from counts of the relative numbers of radio sources of different apparent radio luminosity (or flux density as the radio-astronomers call it). This attempt, made in 1955 by Ryle and P. A. G. Scheuer, is discussed, along with its successors, in chapter 6.

So with this vast catalogue ends the first active period, in which the discovery of a source was more important than any of its detailed properties. However, to discover the 1937th source seemed relatively less important, and so radio astronomers turned their attention to the problem of obtaining more reliable positions for the known sources and of increasing the accuracy of the catalogues, which were suspected of containing spurious sources.

As an example of this correcting process let us consider a survey made by Mills and Slee in 1957. They chose an area of sky which overlapped with that of the 2C survey, and found that in the region of overlap there were many 2C sources that they could not detect. Equally they could detect many sources that were not in the 2C catalogue. The reason for these discrepancies turned out to be that both surveys were confusion-limited. The resolving power of a radio telescope, like that of an optical telescope, is roughly given by λ/a where λ is the wavelength of the radiation being detected, and a is the aperture of the telescope. This means that two sources whose angular separation in the sky is less than about λ/a radians would not be detected as two separate sources. Since radio wavelengths are very much longer than optical wavelengths, a radio telescope has to be very much larger than an optical telescope to achieve the same resolution. Only recently in Ryle's one-mile aperture synthesis system and in long baseline interferometers have similar resolutions been achieved.

The trouble with the 2C survey was that towards its limiting flux density there are so many sources in the sky that many of them came too close together to be properly resolved, resulting in spurious sources being recorded. Only at high flux densities is this difficulty unimportant, and the size of the catalogue then drops considerably. Accordingly the Cambridge radio-astronomers made a new survey at about half the wavelength of 2C, thereby gaining in resolution. This survey, published in 1959, is known as 3C and contains 471 sources.

Further work showed that this survey still had significant defects and in 1962 A. S. Bennett published a revised 3C catalogue that appears to have very few errors in it indeed. This catalogue is of capital importance in the history of radio-astronomy. For many years it provided the main list of radio objects in the northern hemisphere for further radio and for optical study, as the number of well-known objects with 3C numbers adequately testifies. Only now are its successors, the 4C and 5C surveys, beginning to take over. It is therefore worth stating that the revised 3C catalogue gives a virtually complete list of the sources between declinations −5° and +90° that are brighter than 9 flux units* at 178 megahertz. These sources number 328.

During this period the Australian workers were also very active. Between 1958 and 1961 Mills, Slee and E. R. Hill published an extensive catalogue. More recently the great 210-foot paraboloidal dish in Parkes, Australia has come into operation and under Bolton's leadership has produced a major survey. This survey is as important for the southern hemisphere as the Cambridge surveys are for the northern.

Thus in the last 20 years we have moved from the discovery of the first radio source to what may seem to the reader to be an endless catalogue of catalogues. In fact many more such catalogues now exist and further ones are in preparation. They have put the radio-astronomer in the same position as the optical astronomer at the time of the Herschels when star catalogues formed the basis of the study of the Galaxy. As we shall now see, however, the radio catalogues have contributed more to the study of the Universe outside our Galaxy than to our Galaxy itself.

* 1 flux unit is 10^{-26} watts per square metre per hertz.

THE OPTICAL IDENTIFICATION OF RADIO SOURCES

The early radio-astronomers were in the same position as the early optical astronomers for another important reason. They did not know how far away their sources were. They could not even regard the Sun as a *typical* radio source (although it is in fact a radio source) whereas it is a typical star. As with the stars, early parallax measurements could put only a rather small lower limit on the distances of the radio sources. For instance, Ryle and Smith showed in this way that Cygnus A and Cassiopeia A are further than 2×10^{16} cm, which places them outside the solar system but not necessarily as far as the nearest star. The best hope of making further progress lay in attempting to find an optical counterpart to each radio source, since optical astronomers have developed many methods of distance determination. Of course, success could not be guaranteed in advance. Even now the majority of sources used in statistical studies, such as the number–flux density relationship (chapter 6), have not been optically identified.

It is not sufficient for an identification that the optical counterpart be bright enough to be photographed by a large optical telescope. The critical quantity is the uncertainty in the radio position. In the early days the error rectangle contained a large number of optical objects and identification was very difficult. The hope at that time was that the optical object would look abnormal in some way, perhaps as a manifestation of the disturbances that gave rise to the radio source. The chance of finding an abnormal object in the error rectangle by coincidence might well be very small.

For this reason the earliest identifications did tend to be of abnormal objects, but it by no means followed that most optical counterparts of radio sources would be abnormal. Indeed as radio positions improved it soon emerged that most of the optical objects were not particularly abnormal. This improving of radio positions has been continuing up to the present day, and is a vital development because once all the optically brightest sources had been identified it became necessary to seek identifications with optically fainter objects. These objects are more crowded together in the sky than the bright ones, and so a smaller error rectangle is needed to rule out a chance coincidence.

The first people to propose optical identifications were Bolton, Stanley and Slee in 1949. They tentatively identified Taurus A with the Crab nebula, Virgo A with the galaxy NGC 4486, and Centaurus A with the galaxy NGC 5128. These identifications have since been amply confirmed. All the objects concerned are of the greatest interest, and their radio and optical appearance are shown in plate 9.* The Crab nebula was discussed in chapters 1 and 2. It is a supernova remnant and a prime manifestation of the relation between a violent explosion and a strong radio source. As we saw, its radio and optical polarisation are good evidence that it is permeated by a magnetic field in which relativistic electrons are orbiting and producing radio and optical radiation by the synchrotron process. This relationship is believed to characterise radio sources in general, although the intensity of the polarised optical radiation may be too low to be detected. It can be detected in Virgo A, however. The light in the jet coming out from its nucleus which can be seen in plate 10 is polarised. It is also interesting that Arp has found in 1967 evidence for a faint continuation of the jet in the opposite direction. Finally Centaurus A is of interest because, as the radio contours of plate 9(c) show, it is a multiple source which suggests that the explosions may be recurrent. It was also, incidentally, the first extragalactic radio source in which polarisation of the radio emission was found.

These identifications were followed in 1950 by the discovery of Ryle, Smith and Elsmore that the Andromeda nebula (plate 11) and several other nearby galaxies are radio sources. These sources have an intrinsic radio power comparable with that of the Milky Way. When integrated over all radio wavelengths, this amounts to about 10^{38} erg s^{-1}, which is small compared to the optical power of about 10^{44} erg s^{-1}. The situation is very different for what came to be called the radio galaxies, as we shall now see.

The key year here is 1954 when Baade and R. Minkowski identified Cygnus A with the brightest member of a faint cluster of galaxies (plate 12). Up till that time some radio astronomers took

* (*Note added in proof.*) All these objects are now known to be X-ray sources.

the view that the majority of radio sources lay within our Galaxy (the Virgo A and Centaurus A identifications being still tentative at the time). Indeed it was widely held that the background radio emission from the Milky Way resulted from the integrated radiation from these radio sources. The hypothesis that this background came from interstellar space, via the synchrotron mechanism, only gradually gained the upper hand in those years (except in Russia where it was accepted by some physicists such as Ginzburg from the first).

With the aid of more accurate radio positions Baade and Minkowski confirmed the identifications of Taurus A, Virgo A and Centaurus A, and identified Cygnus A as mentioned above. This last identification was based on a good position due to Smith which had a precision of one second of time in right ascension and one minute of arc in declination. Its great importance historically stems from the fact that it was the first of the really powerful radio sources to be identified. The way this was done has become a classical procedure in radio astronomy. Minkowski was able to photograph the optical spectrum of Cygnus A which showed a red shift $\delta\lambda/\lambda$ of 0.057, corresponding to a recession velocity of 17 000 km s^{-1}. On the assumption that this red shift obeys the Hubble law (chapter 3), Minkowski found for the distance of Cygnus A the value 170 megaparsecs. When we recall that Cygnus A is the second most powerful radio source in the sky, we see that its distance is enormous and that the absolute radio luminosity of Cygnus A must be enormous too. It works out at about 10^{45} erg s^{-1} which is 10 million times greater than the radio power of a normal galaxy and 10 times greater than its optical power.

We shall consider later the physical implications of this extraordinary result. Here we must record the unfortunate impact it had on the identification programme. People argued that if the second brightest source in the sky were already 170 megaparsecs away the fainter radio sources would be so distant that they would be too faint to be detected optically even by the 200-inch telescope. This view in one form or another survived until 1960, apparently confirmed by the poor progress of the identification programme up till that year. We now know that this view is incorrect. Only the most powerful radio sources are as strong as Cygnus A. The reason for the paucity of identifications was simply that the radio positions were not good enough.

Thus by about the end of 1960 fewer than 10 per cent of the sources in the 3C and Mills, Slee, Hill catalogues had even tentative identifications. Such as they were these identifications showed that no spiral galaxies were amongst the strong radio sources. These strong sources were all elliptical or SO galaxies, and had a large absolute optical luminosity. To emphasise their great radio power they are called radio galaxies. By contrast weak radio sources like our own Galaxy and Andromeda are called normal galaxies. It was found that about 30 per cent of the radio galaxies were in clusters of galaxies and were then the optically brightest member of each cluster. This proportion still holds good to-day.

Except for the optically brightest sources a positive identification usually requires a radio position accurate to at least 15 seconds of arc in each co-ordinate. This accuracy was first achieved by R. B. Read in 1963 working at the California Institute of Technology. He obtained positions for 110 sources, mostly from the 3C catalogue, to an average precision of 13 seconds of arc. To-day accurate positions are known for many 4C and Parkes sources as well as those in other catalogues, and several identification programmes are under way.

By far the most important result of this identification work is the discovery of a new class of radio source – the quasi-stellar objects (QSOs) whose existence was first clearly established in 1963. The story of this great discovery is told in the next chapter. For the rest it is a question of a rapid increase in the number of radio galaxies identified. In particular the sources in the revised 3C catalogue have by now been largely identified. One obstinate remaining puzzle is that of the blank fields corresponding to sources with good radio positions for which no optical object is visible in the error rectangle down to the limits of the photographic plate. There is still no agreement on whether these blank fields contain mainly radio galaxies or QSOs.

PHYSICAL PROPERTIES OF RADIO GALAXIES

Many observations have now been made of the detailed properties of radio galaxies. Despite all these data we understand very little of their structure and even less of their origin. It would not be appropriate therefore to give a detailed account of these questions here. A few words should indicate the nature of what is involved. We shall consider in turn their angular diameters, or more generally, their radio brightness distribution, their radio spectra, polarisation and optical spectra.

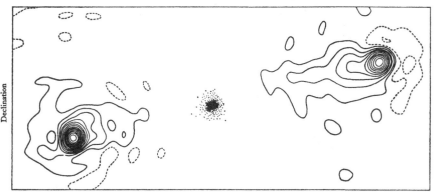

Right ascension

Fig. 24. Contours of the radio galaxy Cygnus A (after S. Mitton and M. Ryle, *Monthly Notices of the Royal Astronomical Society* **146**, 223 (1969)) superposed on a photograph showing the optical galaxy midway between the two radio sources.

The major fact about their brightness distribution is that most radio sources are double. The first example of this was Cygnus A which was resolved in 1956 into two components 85 seconds of arc apart symmetrically disposed on either side of the associated optical galaxy (fig. 24). In some radio galaxies these components are enormous, comparable in size with a whole cluster of galaxies. Such a configuration strongly suggests that a violent explosion occurred in the optical galaxy sending out in two opposite directions a stream of ionised gas (plasma) and relativistic particles. Even granted the explosion the subsequent behaviour of the system is not understood. There are cases like that of Centaurus A (plate 9(c)) whose multiple structure suggests that repeated explosions can occur. Another remarkable multiple source has been discovered by the Cambridge radio-astronomers. Its structure at 1407 MHz is shown in fig. 25.

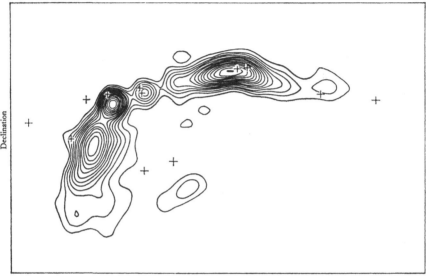

Right ascension

Fig. 25. A remarkable multiple radio source: 3C 465 at 1407 MHz. The crosses represent optical galaxies. (After G. H. Macdonald, A. C. Neville and M. Ryle, *Nature* **211**, 1241 (1966).)

All this evidence for the existence of large scale explosions reminds one of the supernova explosions, which until the discovery of radio galaxies were regarded as the most powerful in the Universe. Explosions on an intermediate scale are also now known or suspected. A good example is provided by the so-called Seyfert galaxies, which C. K. Seyfert originally discussed in 1943, and to which attention was drawn in the present connexion by G. R. Burbidge, E. M. Burbidge and Sandage in 1963. The nuclei of these galaxies appear to have exploded violently some time in the past.* The evidence for this is that the optical spectra contain

* (*Note added in proof.*) Some Seyfert galaxies are now known to be X-ray sources.

emission features not normally seen in the spectra of galaxies, indicating a very high degree of ionisation. In addition the hydrogen lines have very large widths, which, if interpreted as Doppler motions, correspond to velocities of up to 4000 km s^{-1}.

The most spectacular known example of an explosion which is still going on is the radio source 3C 231 which has been identified with the nearby peculiar optical galaxy M 82. A photograph taken in 1963 by Sandage in the light of the hydrogen line H$_\alpha$ shows the explosion (plate 13). Large-scale filaments of hydrogen along the minor axis from the central regions are clearly visible. These filaments are highly polarised.* An analysis of intensities and velocities show that over 5 million solar masses of hydrogen were expelled from the nucleus about a million years ago. The kinetic energy of the moving gas is about 10^{55} ergs. The optical appearance of this explosion is very impressive but we shall soon see that in the most powerful radio galaxies we have to do with very much larger explosive energies.

As regards the radio spectra we need only say that typically they are much like the spectrum of our own Galaxy and so are consistent with the generally accepted hypothesis that the radio emission is due to the synchrotron mechanism. This hypothesis is further supported by the fact that in many cases this radio emission is polarised.

Finally we consider the optical spectra of those radio sources that have been identified. These spectra give us the critical information of the red shift which, via the Hubble law, tells us their distances. From these distances and the flux densities we obtain the absolute radio luminosity of these sources, and from their angular diameters we obtain their linear diameters. All this is vital information, of course, for studying the structure and energetics of these sources. The red shifts can also be used, in conjunction with the apparent optical luminosity, to compare the Hubble relation for radio galaxies with that for galaxies in general. The results of this comparison have been discussed in chapter 3. Here we simply mention that the largest red shift so far obtained for a radio galaxy is 0.461, for 3C 295. This is far greater than the red shifts found by

* (*Note added in proof*.) This polarisation may arise from the scattering of the light by dust particles rather than directly from the synchrotron process.

Hubble, but we shall see in the next chapter that most known QSOs have much larger red shifts still.

THE ENERGY PROBLEM

The staggering nature of the energy problem in radio galaxies was first pointed out by G. R. Burbidge in 1956. He made a now-famous estimate of the minimum energy involved by reasoning as follows. Assuming that a source radiates by the synchrotron mechanism, there are at least two contributions to the total energy stored in the source, namely, the magnetic energy and the energy in the radiating electrons. Initially we know neither the magnetic field strength nor the electron flux. However, the lower the magnetic field the greater the electron flux required to produce the observed emission. It turns out that the total energy is a minimum when there are about equal amounts of energy in the magnetic field and the electrons. This minimum energy is usually enormous. In Hercules A, for instance, it is about 10^{60} ergs.

This may be a considerable underestimate for the following reasons:

(i) So far as we know, equipartition of energy between field and particles need not obtain.

(ii) There may be present a flux of cosmic ray protons whose radio emission is negligible. In the Milky Way, for instance (admittedly a much weaker radio source than what we are calling radio galaxies), there is about 30 times more energy in cosmic ray protons than in electrons.

(iii) The explosion is unlikely to be nearly 100 per cent efficient in converting the explosive energy just into the form in which it can be readily detected, that is, into magnetic fields and relativistic electrons.

It is not possible at present to do more than simply guess how important these factors are. An increase in the minimum energy by a factor lying between 10 and 100 is considered conservative by some astrophysicists. On the other hand it may be possible to reduce the estimates by taking the magnetic field to be very irregular in structure (in agreement with the trend of recent observations).

If the main emitting volume is smaller, for a given power-level, then the total energy turns out to be smaller.

Let us compromise and suppose that the explosive energy in Hercules A is about 10^{60} ergs, our original estimate. Now this is the energy contained in the *rest-mass* of a million solar masses. If we were dealing with matter–antimatter annihilation (as has, indeed been suggested) then the whole of the rest-mass of the material involved might be turned into explosive energy. With other mechanisms we should expect only about 1 per cent at most of the rest-mass energy involved to be released. Thus we are probably faced with the problem of understanding how 10^8 solar masses, one part in a thousand of a whole galaxy, can co-operatively indulge in a violent explosion, in some cases repeating the process on a time-scale, to judge from Centaurus A, of perhaps 10^7 years.

This is perhaps the main physical problem that we face in astrophysics at the present time. There has been no shortage of suggested solutions. The trouble is that all of them are inevitably complicated, and it is difficult to subject any of them to rigorous analysis. Accordingly it is difficult to make specific predictions that would enable some of the proposals to be ruled out. Most of them can be modified, and reasonably modified, to fit a variety of observational facts. For this reason we shall do no more than list some of them:

(i) Matter–antimatter annihilation.

(ii) Catastrophic gravitational collapse followed by expansion (and perhaps including significant nuclear reactions).

(iii) Rapid collisions in an assembly of closely packed stars.

(iv) A chain reaction of supernova explosions.

(v) Hydromagneto–gravitational instability (a large-scale version of a process proposed to account for solar flares).

(vi) Instabilities associated with rapidly rotating highly magnetised bodies (in analogy with the rotating neutron star models of pulsars).

This fascinating problem is unlikely to yield to a stroke of genius. It will probably require for its solution the slow accumulation of understanding that would follow a careful study of all the alternatives. In this unspectacular way we may hope eventually to understand one of the most spectacular of all natural phenomena.

5 Quasi-stellar objects

One of the difficulties in astronomy is that descriptive names are often given to a class of objects before their true nature is understood. Naturally enough these names often turn out to be misleading and this can sometimes constitute a real barrier to clear thinking. Call an object a 'radio star' for long enough (as was done in the early days of radio-astronomy) and it becomes difficult to remember that it may not be a star at all, in any reasonable sense of that word (in fact most of the sources in radio catalogues are radio galaxies). Something of this semantic difficulty surrounds the naming of the objects to be described in this chapter. They were originally called quasi-stellar radio sources or quasars for short (a name especially deprecated by the Californian astronomers, but which has become widely adopted). Then Sandage discovered what he called quasi-stellar galaxies (QSGs) (also called radio-quiet quasi-stellar sources, blue stellar objects (BSOs) and inter-lopers). As we shall see, their properties are similar to those of quasars in all respects except in the relative weakness of their radio emission. We shall therefore follow the Burbidges in giving both types of source the one name quasi-stellar objects (QSOs). Even this name suggests a closer relation to real stars than actually exists, but it does not seem to be worth inventing a totally new name.

THE DISCOVERY OF QSOs

The history of QSOs is even shorter than that of radio galaxies, but much more confusing. In some cases results were fully published some years after they had been briefly announced at conferences.

In other cases similar proposals were made by several people independently at much the same time, and it is often difficult to point to the exact later moment when such proposals were sufficiently confirmed to be regarded as established. Finally, after a slow beginning the later developments became extremely rapid. It is not surprising then that such surveys as exist do not agree with each other in complete detail. We have not attempted a critical historical analysis, and we apologise to all those whose work has been unwittingly slighted.

The story appears to begin in 1960, at which time angular diameters were known for the brightest 3C sources, thanks mainly to the work at Jodrell Bank. Several of these radio sources had remarkably small angular diameters and so were of special interest. It later turned out that this was a somewhat accidental approach to the discovery of QSOs, many of which in fact have substantial radio angular diameters. At any rate in 1960 it was considered to be of great interest that 3C 48, 3C 286, 3C 196 and 3C 147 had unusually small angular diameters. In September of that year Sandage took photographs with the 200-inch telescope of the regions containing the first three of these sources. These photographs were studied by T. A. Matthews who found that in each case the only object in the error rectangle was what appeared to be a star (plate 15). In October Sandage obtained a spectrum and photoelectric colours for the star close to 3C 48, and announced his results at the December meeting of the American Astronomical Society. However, a detailed account of this work was not published until 1963.

The optical spectrum of 3C 48 was very strange. It consisted of broad emission lines which could not be identified. The colour of 3C 48 was also somewhat strange but not unprecedentedly so, being similar to that of white dwarfs, old novae, and irregular variables of the U Geminorum type. The colour is obtained by measuring the brightness of the object in three different wave-length ranges called U, B, V (ultra-violet, blue, visible) selected by means of filters with appropriate wavelength responses (fig. 26). With the U, B, V intensities measured logarithmically in terms of magnitudes, the colour can be represented by the differences

$U-B$ and $B-V$, which then give intensity ratios. Fig. 27(a) shows graphically the colours of main sequence stars, with their values of $U-B$ and $B-V$ plotted against each other. By contrast, white dwarfs, old novae and irregular variables occupy a different part of such a diagram, as shown in fig. 27(b). The physical reason for this difference is that the main sequence stars are cooler than the others. The hydrogen in their outer layers is then un-

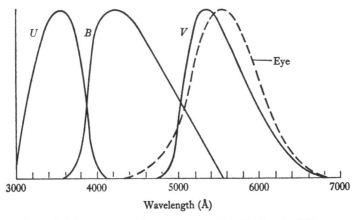

Fig. 26. The wavelength responses of the U, B and V filters, and of the eye.

excited and able to absorb some of the ultra-violet radiation emitted from the interior. In the hotter stars most of the electrons in the hydrogen atoms are excited to energy levels well above the lowest, from which they absorb very little ultra-violet radiation. Stars in the white-dwarf region are usually described as having an ultra-violet excess, although it would be more logical to say that main sequence stars have an ultra-violet deficit.

Since 3C 48 lies fairly close to the white-dwarf region in the $U-B$, $B-V$ diagram we can conclude that its surface is hot, but very little else. The real mystery lay in the emission spectrum (plate 16) whose lines could not be identified. As late as December 1962 attempts were being made to explain it in terms of anomalously high excitation leading to the production of unfamiliar lines. At that time 3C 48 was thought to be a star in our own Galaxy,

perhaps 100 parsecs away. This view was based not only on its
starlike appearance (being unresolved by the 200-inch telescope),
but also because the photoelectric data showed that the optical

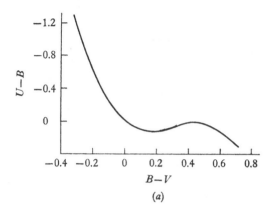

(a)

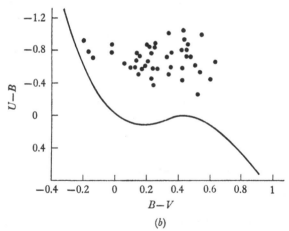

(b)

Fig. 27. (a) Two-colour diagram for main sequence stars.
  (b) Two-colour diagram for white dwarfs, old novae and irregular variables, showing their 'ultra-violet excess' relative to main
sequence stars. QSOs have a similar ultra-violet excess.

brightness of 3C 48 varied appreciably on a time-scale comparable
with one day. It was argued that light could not take longer than
this time-scale to cross the varying region or the variations would
be smeared out. This region could not then be larger than one

light-day, and might of course, be very much less, a conclusion which supported the view that the object was a star rather than a distant galaxy.

All this was changed early in 1963, a climacteric year for astronomy and cosmology. The first step was actually taken in the autumn of 1962, but an account of it was first published in 1963. It consisted of the identification of another 3C source, 3C 273, with altogether unprecedented reliability. This was made possible by a positional determination which had an uncertainty of less than 1 second of arc. This remarkable achievement we owe to Hazard, Mackey and Shimmins who used the Parkes telescope to observe a lunar occultation of 3C 273. By great good fortune the moon, as seen from Parkes, passed three times across 3C 273 in 1962. Since the position of the edge of the moon at any time is known very accurately, a careful timing of the disappearance and reappearance of the occulted source provides a very accurate position for it. Nor is this all. The source does not disappear and reappear with complete abruptness. In fact the edge of the moon acts as a diffracting screen, and a diffraction pattern is obtained. From the observed pattern one can estimate the size of the source and something about its detailed structure.

In the present case a beautiful diffraction pattern was obtained (fig. 28), not much inferior to the patterns that are used as illustrations in text-books on optics. Hazard, Mackey and Shimmins found that the radio source consisted of two components A and B 20 seconds of arc apart (fig. 28). The smaller of these two components 3C 273 B (which remains partially unresolved to this day) coincided in position with a 13th magnitude* blue star (plate 17). The agreement in position was so good that the identification was accepted as completely reliable. This coincidence in position was rather fortunate because, as we saw in the last chapter, the radio and optical emissions from many sources come from different regions, and this is true of some QSOs as well as of radio galaxies.

The successful identification prompted M. Schmidt to obtain an optical spectrum of 3C 273 B (plate 18). At the same time he

---

* The magnitude scale is defined by $m = -2.5 \log$ brightness. On this scale the Sun has an apparent magnitude of about -27.

discovered a faint bluish jet extending from the B component to the A component (plate 17). The spectrum was similar to that of 3C 48 in that it consisted of broad emission lines which could not at first be identified. Then came the historic moment. Schmidt decided to see whether he could interpret the spectrum in terms of a substantial red shift despite the presumption that the object was a star in our Galaxy. He was immediately successful. Four of the emission lines fitted very well a standard series of hydrogen lines with a red shift $\delta\lambda/\lambda$ of 0.158 (a fifth line in the red being dis-

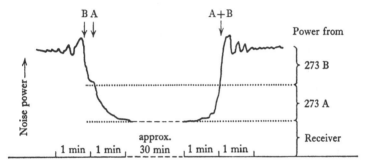

Fig. 28. Radio emission from the QSO 3C 273 measured during its occultation by the moon on 5 August 1962. (Reproduced by kind permission of C. Hazard.)

covered later by J. B. Oke), while the other emission lines also had immediate interpretations in terms of this red shift. If this is a Doppler shift the 'star' is moving away from us with nearly 16 per cent of the speed of light. This result was published early in 1963, and the era of the QSO had begun.

Developments in several directions followed rapidly. It was especially attractive to study 3C 273 B further, because being relatively bright it could be found on photographic plates taken many years earlier for other purposes. It was in fact traced back for 70 years on plates taken at the Harvard and Pulkova Observatories. This old material was useful for studying light variations on a longer time-scale than Sandage could achieve.

It was immediately evident that if the red shift of 3C 273 B obeyed the Hubble law, as the red shifts of radio galaxies appear to do, then this source must be exceedingly bright in absolute optical power. For its distance would be about 500 megaparsecs, and since its apparent brightness is of 13th magnitude, its absolute brightness must then be about 100 times greater than that of the brightest known galaxy. Leaving aside for the moment the startling nature of this result and the physical problems it raises, we see that it would imply that other QSOs at much greater distances should still be readily detectable and yet have very large red shifts indeed. A first step towards the realisation of this was achieved almost immediately. Stimulated by Schmidt's discovery, J. L. Greenstein and Matthews solved the mystery of the spectrum of 3C 48. This source is 3 magnitudes fainter than 3C 273 B, and its spectrum becomes readily understood if it has a red shift of 0.367. This result was also published in 1963.

1964 was mainly a year of consolidation. Eight more QSOs were identified and Schmidt and Matthews obtained a red shift of 0.545 for 3C 147, thereby beating the record for a radio galaxy (0.461 for 3C 295). The most important development was the introduction by Ryle and Sandage of a rapid technique for finding QSOs. This exploited the fact that QSOs appeared to have an ultra-violet excess. It consisted of taking two plates in the general vicinity of good radio positions, one through the U filter and one through the B filter. 'Stars' which were brighter through the U filter could then be rapidly picked out. In this way Ryle and Sandage identified 3C 9, 3C 216 and 3C 245 as QSOs.

The year 1965 saw several dramatic developments. Not surprisingly there was a rapid increase in the number of QSOs known. In addition the individual properties of QSOs began to be studied in detail. The most remarkable of these properties arose from the measurement of the infra-red emission of 3C 273 B at around 10 microns (1 micron is 10^{-4} cm) by F. J. Low and H. L. Johnson. The spectrum of 3C 273 from the radio via the infra-red to the optical and (recently) the X-ray region is shown in fig. 29. It can be seen from this spectrum that 3C 273 B emits more energy at infra-red wavelengths than at all other wavelengths combined, a very unusual state of affairs (since found to obtain for some Seyfert galaxies also).

The greatest single development in 1965 was Schmidt's discovery of large red shifts, although it was not altogether to be unexpected, as we saw above. Determining large red shifts is not easy, partly because the spectral lines are often so faint that it takes an experienced eye to distinguish them from graininess in the photographic plate. (Later work employing image tubes has eased

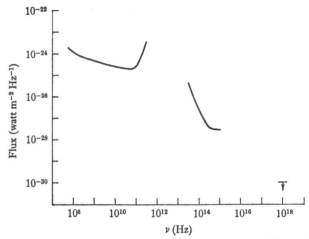

Fig. 29. The continuous energy spectrum of 3C 273. There is now some convincing evidence that the actual X-ray flux is fairly close to the upper limit shown at 10^{18} Hz.

this difficulty.) Another problem is that large red shifts bring the ultra-violet region into the visible, and since this is an unfamiliar region, being inaccessible to ground-based instruments, one does not know what lines to expect. The correct identification of lines is accordingly a tricky business. Schmidt took advantage of the fact that the lines already identified in QSOs had also been found in nebulosities called planetary nebulae in our own Galaxy. Assuming that the physical conditions in QSOs where the lines are formed are similar to the fairly well-known conditions in planetary nebulae, Schmidt drew up a search list of ultra-violet lines based on calculations by D. E. Osterbrock for the ultra-violet spectra of the nebulae. In using this list Schmidt was greatly helped by the fact that he found several red shifts intermediate between the previously existing ones and the very largest. A spectrum with an intermediate red shift would contain lines at the red end already found in QSOs with small red shift, and some new lines at the blue end. For instance 3C 287 with a red shift of 1.005 had the lines of singly ionised magnesium Mg^{II} (rest wavelength 2798 Å), doubly ionised carbon C^{III} (rest wavelength 1906 Å), and C^{IV} (1550 Å). Having thus established that the C^{IV} line appears, as would indeed be expected on the basis of Osterbrock's calculations, Schmidt was

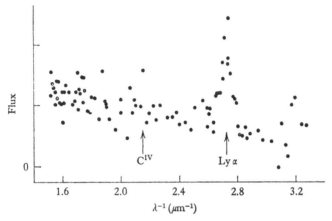

Fig. 30. The photoelectric spectrum of the QSO 3C 9. It has a red shift of 2.012, which brings the Lyman α line (1216 Å) right into the visible at 3666 Å. (After E. J. Wampler, *Astrophys. J.* **147**, 1 (1967).)

able to interpret spectra with larger red shifts in which the C^{IV} line would be at the red end. By this step-by-step process Schmidt found a series of large red shifts culminating in the fantastic shift of 2.012 in 3C 9 (fig. 30), a source in which, for the first time, the basic hydrogen line Lyman α (1216 Å)* was seen from the ground, shifted into the visible at 3666 Å. This great result required some intricate argumentation to justify, but so many large red shifts are now known† that there is no longer any spectroscopic doubt about the interpretation.

---

* This line is emitted when an electron in a hydrogen atom drops from the first excited state into the ground state.
† The record at the time of this writing is a red shift of 2.88.

A red shift of 2.012 is so large that the normal notation for red shift becomes misleading. In Hubble's day when red shifts were very much less than 1, it was customary to write

$$z = \frac{\delta\lambda}{\lambda}$$

for the red shift. Now, with $\delta\lambda$ sometimes twice as big as λ, it is better to write $1 + z$ for the factor by which the wavelength is increased. In 3C 9 this factor is 3.012, which brings the Lyman α line right into the visible. We can also no longer write with Hubble

$$z = \frac{v}{c},$$

which is the classical Doppler shift relation. If for the moment we ignore cosmological subtleties, and simply regard the red shift as a special relativistic Doppler effect (other interpretations will be mentioned later) then we must write

$$1 + z = \sqrt{\left(\frac{c+v}{c-v}\right)},$$

a relation that is illustrated in fig. 31. The red shift clearly tends to infinity as the recession velocity v approaches c. For 3C 9, v works out to be about 80 per cent of c.

The significance of this result will occupy us in many places in this book. Now, however, we must consider another important discovery of 1965, Sandage's quasi-stellar galaxies. This discovery arose from further applications of the Ryle–Sandage two-colour method of finding QSOs at or near radio positions from the ultra-violet excess. It was found by Sandage and C. R. Lynds that many ultra-violet objects turned up which were nowhere near the radio positions. At first they were regarded as nuisances – Sandage called them interlopers at the time. Fortunately he had second thoughts and began to study them intensively. It soon emerged that they were similar to the blue stars previously studied by G. Haro and W. J. Luyten. For various reasons he came to suspect that many of them were not stars at all, but were radio-quiet quasi-stellar sources. Curiously enough his reasons turned out to be misleading (so we shall not give them here) but his conclusion was correct. This became clear after optical spectra were taken of six of them. Sandage found that one was a star, two had continuous spectra with no emission or absorption features, and three were extragalactic. Of these, one had a non-stellar image and a small red shift, but the other two had completely stellar images. These

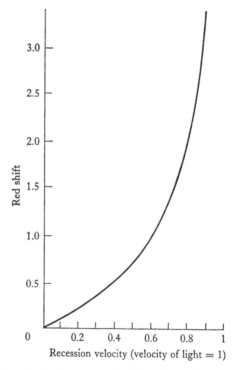

Fig. 31. The relation between red shift and velocity, according to special relativity. For small velocities the relation is the classical, linear one. As the velocity approaches that of light the red shift grows without limit.

two had optical spectra like those of the radio QSOs. One of them, Ton 256 (Ton = Tonanzintla) had a red shift of 0.131. The other, BSO 1 (BSO = blue stellar object) had a moderately large red shift by the new standards set by Schmidt's results, namely, 1.242.

By virtue of his indirect arguments Sandage estimated that about 80 per cent of the fainter blue 'stars' were extragalactic. However, T. D. Kinman, and Lynds and G. Villere were able to show that this was an overestimate by about a factor of 5. Even so there would be nearly 100 times more of these radio-quiet objects than of the original quasi-stellar radio sources. Sandage's discovery is therefore of great importance. However, it must be emphasised that the radio-quietness of these new objects is only relative. Recently several coincidences have been found between blue stellar objects and weak 4C sources. In addition at least one BSO, namely PHL 1222 (PHL = Palomar, Haro, Luyten) has been individually detected as a very weak radio source.

The situation appears to be that if a set of objects is selected by optical characteristics (blue colour and stellar image) their absolute radio luminosity varies over a wide range, with more objects at the fainter levels. This is perfectly reasonable and indeed the indications are that a range of about 100 to 1 in absolute radio power level suffices to include most of the extragalactic blue 'stars' and quasars. This is very much smaller than the corresponding range for radio galaxies, which is more like a factor of 10^6. It therefore seems very likely that quasars and QSGs belong to the same population of objects, their different methods of discovery merely reflecting the state of the radio art at the time. For this reason we shall follow the Burbidges and call both sets of objects QSOs.

The last of the great discoveries about QSOs made in 1965 was that some of them are *radio* variables as well as optical variables. This may not seem a very exciting result by the new standards we have been set, but it caused a sensation at the time because it seemed at first sight to mean that the varying QSOs could not be at the great distances implied by the Hubble law. The cosmological significance of the QSOs thus hung in the balance. Even now this question is not completely settled. The story begins with a discovery that is no longer believed to be correct. (The reader has by now probably become accustomed to the reversals of fortune so common in astronomy.) In February 1965 the Russian radio-astronomer G. B. Sholomitsky announced that the QSO CTA 102 (CTA = Cal. Tech. A) varies by about 30 per cent at 940 MHz, the period being about 100 days. Now for a reason we shall discuss in a moment it was believed at that time that the angular diameter of

CTA 102 was at least one-hundredth of a second of arc. Since the varying part of the source could hardly be much larger than 100 light-days (or the variations in the source would be smeared out), Sholomitsky concluded that it must be closer than 2 megaparsecs, and might even be inside our Galaxy. However, in April 1965 Schmidt reported that CTA 102 has a red shift of 1.037. If this red shift satisfied the Hubble law the source would be at a distance of about 3000 megaparsecs, that is, 1500 times greater than Sholomitsky's upper limit.

In August doubt was cast on Sholomitsky's result by P. Maltby and A. T. Moffett of the California Institute of Technology. They reported that CTA 102 had not varied appreciably at 970 MHz over a three-year period ending about two years before Sholomitsky's observations began. Similar negative results have since been reported by many sets of observers at many frequencies and during many intervals of time. At the moment it is generally assumed that Sholomitsky's observations were in error.

In the meantime there was a second report of a varying QSO, this time our old friend 3C 273 B. In May 1965 it was reported by W. A. Dent of the University of Michigan to have increased its flux density at 8000 MHz by about 40 per cent in the last three years (fig. 32). This has since been amply confirmed by other workers, and indeed many radio variables are now known. Since the time-scale of the variations of 3C 273 B is a few years, the size of the varying region cannot apparently exceed a parsec or so. On the other hand, the radio angular diameter was thought to exceed about three-hundreths of a second of arc, which would imply that the source must be within about 10 megaparsecs. However, if its red shift of 0.158 is cosmological in origin its distance must be about 470 megaparsecs.

To understand what is involved in this discrepancy we must consider how the angular diameter was estimated. The estimate was based on the fact that if a powerful source is very small its surface brightness must be very large. However, the surface temperature of a body cannot exceed the temperature of its individual radiating elements. As soon as this condition is approached the elements begin to absorb each other's radiation, and we speak of the source as possessing self-absorption. If the QSOs radiate by the synchrotron mechanism, then the presence of self-absorption reveals itself in the radio spectrum of the object, which should drop rapidly at low frequencies. Conversely the absence of a low-frequency cut-off in the spectrum puts an *upper* limit on the surface temperature of the source. For a given flux density this puts a *lower* limit on the angular

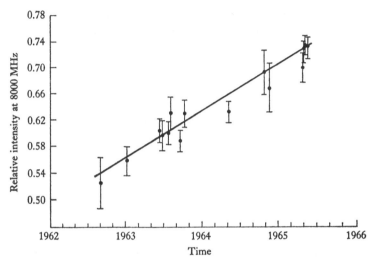

Fig. 32. The radio variation of 3C 273 from 1962 to 1965 – the first well-established evidence for radio variation in a QSO.

diameter. In other words if the same amount of flux had to be squeezed into a smaller angle it would put the surface brightness up to the point where self-absorption would set in.

It is the absence of observable self-absorption in the spectrum of 3C 273 B that yields a lower limit on its angular diameter of about 3 hundredths of a second of arc, a result originally obtained independently by V. I. Slish and P. J. S. Williams in 1963. However if its distance is 470 megaparsecs, the time-scale of the radio variations implies that the angular diameter of the varying region cannot exceed about one-thousandth of a second of arc. Several explanations have been proposed to account for this discrepancy:

(i) The source may be composite. The idea here is that the

varying region would be smaller than one-thousandth of a second of arc and would indeed be self-absorbed. However this self-absorption would not show up in the observed spectrum because it would be masked by the radiation from a larger unvarying part of the source. This explanation may well contain part of the truth because recent observations using an interferometer with a trans-continental baseline have verified that 3C 273 B does indeed contain a component smaller than 10^{-3} seconds of arc. However, it is now known that the source is variable even at the relatively low observing frequency of 1400 MHz. Detailed calculations show that the composite model cannot account for this.

(ii) Accordingly M. J. Rees has suggested that the radio source may be expanding at a speed close to the speed of light. In such a case relativistic effects become important and would permit the varying region to be considerably larger than had been deduced from the time-scale of the variations. The problem of self-absorption would not then arise.

(iii) The source may not be radiating by the ordinary synchro-tron mechanism, which would again invalidate the argument from the absence of observed self-absorption.

(iv) The source may be much closer than the distance deduced from the red shift and the Hubble law. In that case the varying region could have a size determined by the time-scale of the variations without subtending such a small angle at the observer that one is in difficulties over self-absorption. This local hypothesis for QSOs will be discussed later.

By contrast to 1965, 1966 was an uneventful year. Large numbers of QSOs were identified and more red shifts were obtained, but no qualitatively new discoveries were made. The most startling discovery was the optical behaviour of 3C 446. Its brightness varied in a remarkable manner, often by as much as 0.8 of a magnitude (about a factor two) on the time-scale of a day. These are the largest short-period variations yet observed in a QSO. They are fascinating in themselves but they do not raise the same problems as the radio variations because in the optical case there is no ana-logue to the problem of the missing self-absorption. They do however imply small regions and so intense radiation fields, which raise interesting questions about processes like inverse Compton collisions (cf. chapter 15). However, these matters raise structural problems that do not affect the cosmological questions with which this book is mainly concerned, and reluctantly we must abandon them. The most important discovery of 1966 was the presence of absorption lines in the optical spectra of some QSOs. Strictly speaking one absorption line had been found in BSO 1 in 1965, but the first example of a spectrum with many absorption lines was 3C 191, which was studied by the Burbidges, Lynds and A. N. Stockton. There are now many sources known with absorption lines. Attempts to identify these lines have revealed an entirely new problem, namely, that some absorption spectra can be explained only by the hypothesis that several red shifts co-exist in one spectrum. The most obvious interpretation of this is that the QSO has expelled several shells of absorbing material whose relative velocities affect the observed absorption red shifts. In some cases these relative velocities would have to be close to that of light. Yet the absorption lines are often very narrow, implying a small velocity dispersion within each shell. Another peculiarity, stressed especially by Geoffrey Burbidge, is that an absorption red shift of 1.95 occurs far more often than would be expected by chance, as though it were a standard value in relation to some unknown mechanism. Despite some interesting speculations these problems remain unsolved. Moreover since 1966 little that is qualitatively new has been discovered. What we badly need now are more systematic data on QSOs. The obtaining of such data depends critically on the existence of efficient methods of locating QSOs. The two-colour method is very wasteful because it does not discriminate well enough against stars like white dwarfs which heavily dilute the sample. This problem may have been solved by A. Braccesi of Bologna who found that he could eliminate most blue stars rapidly by taking an additional plate in the near infrared. QSOs seem to be much stronger in the infra-red than white dwarfs. This method may provide us with many hundreds of QSOs whose red shifts and other properties could then be examined in detail. In this way the study of QSOs could be put on a proper statistical basis.

PHYSICAL PROPERTIES OF QSOs

We now wish to make some general comments on the situation created by these observations. The first question must naturally be: when do we call an object a QSO? This classification-type problem is often a difficult one in astronomy because there are usually no natural barriers that delimit a particular class of object. In this particular case it has become clear that the radio properties are not the decisive ones. Like radio galaxies, QSOs are sometimes single radio objects, sometimes double, perhaps multiple. They are sometimes strong and sometimes weak, sometimes polarised and sometimes not. On the other hand the optical properties do seem to be decisive, although it is not exactly clear what their defining properties should be. Even the obvious one of a stellar image will not do, since in some cases one can just see some optical structure. Moreover, there is nothing fundamental about being too small to resolve, since this has to do with the size of telescopes as well as with the size and distance of the optical object. Next we might try the ultra-violet excess, but as we shall see in a moment, QSOs with red shifts around 3, if such exist, might not exhibit this excess.

Rather than try to solve this problem (which probably has no solution) we prefer to stress that property of the known QSOs which makes them so valuable to the cosmologist. This is quite simply their enormous absolute optical luminosity transcending, as we have seen, the brightest known galaxies by a factor of up to a hundred. This enables them to be detected and their optical spectra to be studied when they are at much greater distances than the most distant known galaxy. The Universe can thus be explored out to much greater depths, and at depths moreover whose light-time away is comparable to the time-scale of the Universe itself.

The consequences that can be drawn from this will be discussed in later chapters. In preparation for that discussion we consider here two basic questions. The first is, can we expect to detect red shifts substantially bigger than the ones so far detected? The second is, can we be sure that the red shifts have to do with the expansion of the Universe, and so are a direct measure of distance?

The first question has often been raised in the form; why have red shifts of about 3 not yet been detected? The largest shift known to date is for the source 4C 05.34; according to Lynds its red shift is 2.88. The second largest is 5C 2.56 with a red shift of 2.36. There are several other sources whose red shifts exceed 2. These sources are by no means so faint either at radio or optical wavelengths that if they had a red shift of 3 they would be undetectable. The data do rather suggest that there must be a sharp cut-off in the number of QSOs whose red shift exceeds, say, 2.5. Four suggestions have so far been proposed to explain this cut-off:

(i) The first QSOs may have formed during the evolution of the Universe at a time in the past corresponding to a red shift of 2.5.

(ii) QSOs with larger red shifts may exist, but the effect of the large red shift may be to transfer the whole of the ultra-violet excess into the red. So long as an ultra-violet excess is used to select candidates for spectral study, objects with a large red shift would then be missed. If this is the correct explanation, the Braccesi technique of looking for an infra-red excess might enable red shifts greater than 2.5 to be found, although at the moment this technique is very inefficient unless an ultra-violet excess is also used for selecting candidates for spectral study.

(iii) The QSOs of large red shift may be clustered as discussed in chapter 7. In this case the far edges of the clusters so far observed may occur at a red shift of about 2.5, and the next clusters may not begin until a red shift of, say, 3.5. The likely optical appearance of QSOs with such a large red shift is unknown.

(iv) The cut-off may be due to absorption by intergalactic hydrogen, provided that this hydrogen were less highly ionised at early epochs than when $z \leqslant 2$ (see chapters 9 and 10). This suggestion, which is due to Rees, is numerically reasonable, and is the most attractive of the four so far proposed.

We now consider the second question, concerning the origin of the red shifts of QSOs. There are the following possibilities for the red shifts:

(i) They have to do with the expansion of the Universe.

(ii) They are Doppler shifts, resulting from a large peculiar velocity of the QSOs relative to their immediate surroundings. On this view the large peculiar velocity indicates that the QSOs have been thrown out from some relatively local centre or centres by a violent explosion.

(iii) They are gravitational in origin, that is, the light has been emitted by a massive object with a large gravitational potential, leading to a substantial Einstein red shift.

(iv) They arise from a new law of physics.

We have been assuming (i) throughout our discussion, and it is indeed the commonly held view. Nevertheless so much depends on understanding the red shifts that we must consider the other possibilities. It seems too early to take (iv) seriously, although we should mention that Arp claims to have found some evidence to support it. He has analysed the positions of QSOs in relation to peculiar galaxies of quite different red shift. He claims to have found significant correlations, and has suggested tentatively that a new law of physics may be at work. However Arp's statistical discussion has been challenged by other workers and is not generally accepted.

The arguments against (iii), the gravitational red shift, are of two kinds. The first is structural. It is difficult to construct a detailed model of a source that is compatible with all the observations and that leads to the required gravitational red shifts. Arguments of this type should be treated with great reserve. They amount to saying that since we find it difficult to make a suitable model of a certain type, Nature must find it difficult too. This argument neglects the possibility that Nature may be cleverer than we are. It even neglects the possibility that we may be cleverer to-morrow than we are to-day. This is just the situation we are in at the moment as regards the gravitational red shift. There is a straightforward structural argument against it, first devised by Greenstein and Schmidt, and this argument has stood for several years (by QSO standards a very long time). Recently, however, Hoyle and Fowler found a conceivable way around this argument. Theirs is not a particularly plausible model, but who can tell whether to-morrow they or someone else will not be able to improve on it?

There is, however, a non-structural argument against the gravitational red shift, arising from counts of QSOs. As explained in chapter 6, if the QSOs are nearby non-cosmological objects and are nearly at rest, as would be implied by the gravitational hypothesis, then if we are not in a privileged position the number–flux density relationship for them should have a particular form (the three-halves power law) which is not the form observed. (A full discussion of this question is given in chapter 7.)

We come finally to possibility (ii), which supposes that the QSOs have been expelled by a violent explosion in our own or a neighbouring galaxy. This hypothesis was proposed by J. Terrell, and has been supported by the Burbidges and by Hoyle. Structural arguments have been devised both for and against the hypothesis, but for the reasons given above we will not consider these arguments here. The number–flux density relationship for QSOs would no longer be relevant both because we would now be in a privileged position, being close to the centre of the explosion, and because there would be no reason to expect the QSOs to be distributed uniformly. (We return to this point in the next two chapters.)

There are, however, two powerful arguments which show that the local hypothesis requires us to be in a privileged position in a rather unacceptable sense, namely, that only very few galaxies can be currently surrounded by QSOs. Probability arguments in astronomy can be dangerous, but unless it can be shown that there is some relation between QSOs and the development of life, it is an unreasonable consequence of the local hypothesis that we should live in one of those rare galaxies that are surrounded by QSOs.

The first of these arguments is the complete absence of *blue* shifts amongst the QSOs. It would be understandable that the QSOs emitted by our own and perhaps the immediately neighbouring galaxies would by now have moved so far that we see them all receding from us. The QSOs from somewhat more distant galaxies are a different matter, however. We should expect some of them to be moving towards us. In fact since a blue shift would also make a source brighter at both radio and optical wavelengths, it turns out that if sources emitted by a distant galaxy are observationally selected by their apparent brightness (which, of course, in practice they would be) then we should actually see *more* blue shifts than red shifts. The ratio of the expected number of blue shifts to red shifts goes approximately like $(1+z)^3$, a factor which for $z = 2$ is 27. It will not do to argue, as once was done, that a blue shift may be hard to detect because of the paucity of lines in the red and infra-red that would be moved into the visible. There simply are not enough puzzling spectra, or even unidentified strong radio sources, available. It follows that the nearest galaxy whose QSOs are still approaching us must be so far away that all its QSOs are too faint to be observed. Being surrounded by QSOs would then be a rare event for a galaxy.

The second argument leads to the same conclusion. The total radio noise from all QSOs must not exceed the observed extra-galactic radio background. If all galaxies were surrounded by QSOs in the same way as our own is supposed to be according to the local hypothesis, this condition would be broken by a large factor. Again we conclude that a galaxy is rarely surrounded by QSOs.

It seems then that the cosmological hypothesis (i) is the most likely to be correct, and in the rest of this book we shall adopt it.

(*Note added in proof.*) The case for the cosmological hypothesis has recently been strengthened by Gunn's discovery that the QSO PKS 2251 + 11, which lies in the direction of a cluster of galaxies, has the *same* red shift (0.33) as a galaxy in the cluster.

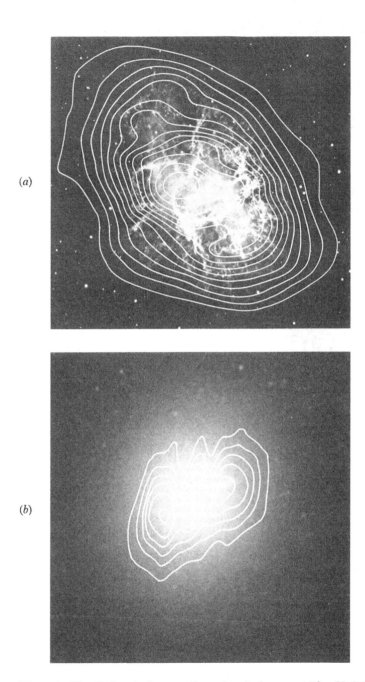

Plate 9 (a) The Crab nebula as a radio and optical source. (After N. J. R. A. Branson, *The Observatory* **85**, 250 (1965).) (b) Virgo A as a radio and optical source. (After I. D. G. Graham, *Monthly Notices of the Royal Astronomical Society* **149**, 319 (1970).)

[*Facing p.* 82]

(i)

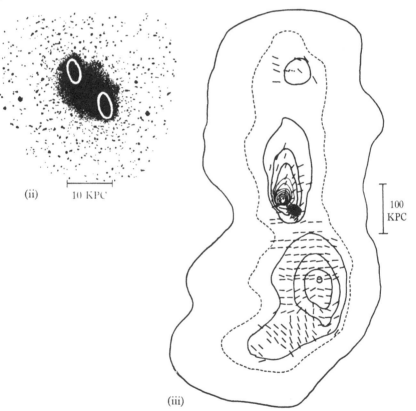

(ii)   10 KPC

(iii)

100
KPC

Plate 9 (c) Centaurus A as a radio and optical source (after a composite
by T. A. Matthews). (i) The optical galaxy. In the radio pictures (ii) and
(iii) it is represented by the central black blob. (ii) The innermost part
of the radio galaxy. It is a double radio source. (iii) The radio galaxy
as a whole. The outer parts form another double radio source. Note the
magnetic field directions derived from polarisation data and indicated
by the bars.

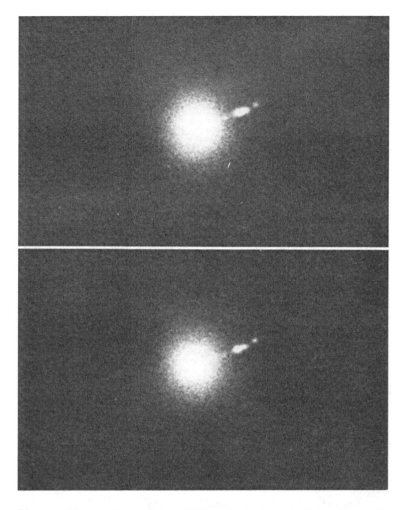

Plate 10. The nucleus and jet of Virgo A observed in light of two different polarisations. (From W. Baade, *Astrophys. J.* **123**, 550 (1956).)

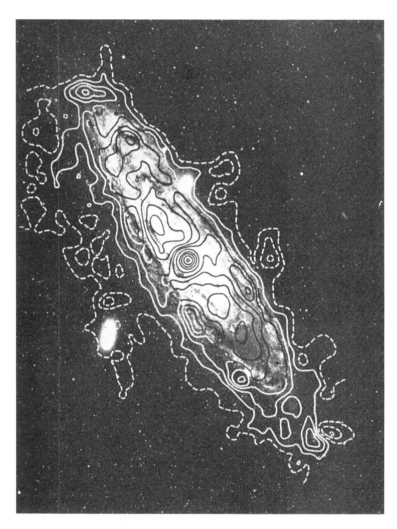

Plate 11. A radio map of the Andromeda galaxy at 408 MHz superposed on an optical picture. (From G. G. Pooley, *Monthly Notices of the Royal Astronomical Society* **144**, 101 (1969).)

Plate 12. An optical picture of the Cygnus A radio source. (Photograph from the Hale Observatories.)

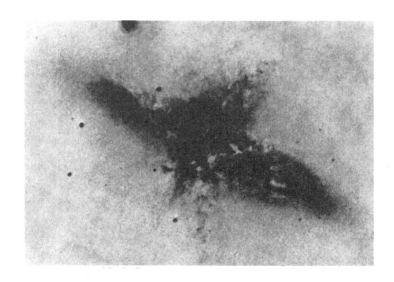

Plate 13. An Hα photograph of M 82 showing the filamentary structure along the minor axis. (From A. Sandage, *Varenna Lectures* (1966).)

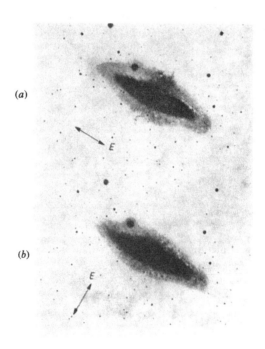

Plate 14. M 82 observed in polarised light: (*a*) With the electric vector predominantly along the major axis. The filaments radiate strongly. (*b*) With the electric vector predominantly along the minor axis. The filaments are very weak. (From A. Sandage, *Varenna Lectures* (1966).)

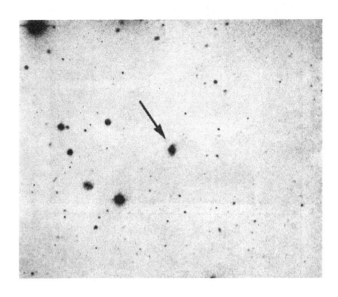

Plate 15. An optical picture of the QSO 3C 48. (From A. Sandage & W. C. Miller, *Astrophys. J.* **144**, 1238 (1966).)

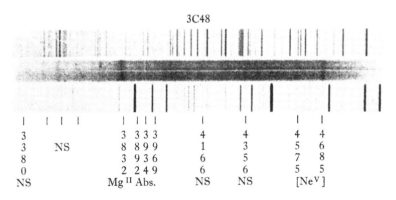

3C48

3		3 3 3	4	4	4	4
3	NS	8 8 9 9	1	3	5	6
8		3 9 3 6	6	5	7	8
0		2 2 4 9	6	6	5	5
NS		Mg II Abs.	NS	NS	[Ne^V]	

Plate 16. Two optical spectra of 3C 48, taken on 12 November and 2ɔ December 1960. (From J. L. Greenstein & M. Schmidt, *Astrophys. J.* **140**, 1 (1964).)

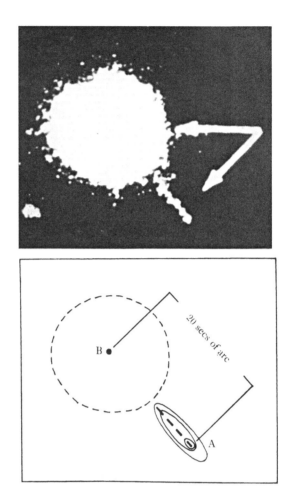

Plate 17. An optical picture of the QSO 3C 273. It is overexposed in order to bring out the very faint jet. (From F. D. Kahn & H. Palmer, *Quasars*, Manchester University Press (1967).)

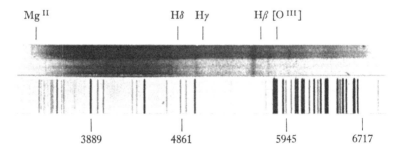

Plate 18. The optical spectrum of 3C 273 B, taken on 23 January 1963. (From J. L. Greenstein & M. Schmidt, *Astrophys. J.* **140**, 1 (1964).)

6 The radio source counts

INTRODUCTION

The first attempt to use counts of radio sources to draw cosmological conclusions was made by Ryle and Scheuer in 1955. They used the 2C catalogue described in chapter 4, and came to the conclusion that the counts were incompatible with the steady state theory of Bondi, Gold and Hoyle with its continual creation of matter (p. 116). This claim gave rise to considerable controversy, echoes of which can still be heard occasionally to-day. We shall try in this chapter to give an unbiased account of the present situation, but the reader is warned that the phrase 'radio source counts' still seems to provoke stronger feelings in radio-astronomers than any other phrase except perhaps 'local hypothesis for quasars'. Accordingly any expert who might happen to read this chapter will probably think that it is either too strong or too weak.

The counts themselves consist of the number $N(S)$ of radio sources per unit solid angle whose measured flux density at the operating frequency of the radio telescope exceeds the quantity S. As we shall see, the relation between N and S which would be expected for a uniform distribution of stationary sources has the form

$$N \propto S^{-\frac{3}{2}}.$$

A plot of $\log N$ against $\log S$ would then be expected to be a straight line of slope $-\frac{3}{2}$. As we shall also see, when the red shift is taken into account the quantity $NS^{\frac{3}{2}}$, instead of being independent of S, should decrease with decreasing S. In other words, the $\log N$–$\log S$ curve should be *flatter* than in the static case. The observed curve is, however, *steeper*.

THE OBSERVED COUNTS

The anomalous steepness found by Ryle and Scheuer was very marked indeed. For the fainter sources in their analysis the slope of the $\log N$–$\log S$ curve was -3. However we have seen that the 2C survey was confusion-limited, which means that many of the

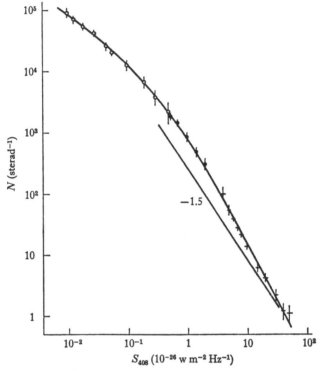

Fig. 33. Counts of radio sources derived by Pooley and Ryle. N is the number of sources per unit solid angle whose flux density at 408 MHz exceeds S_{408}. (From G. G. Pooley and M. Ryle, *Monthly Notices of the Royal Astronomical Society* **139**, 515 (1968).)

faint sources that were recorded were actually spurious. Some part at least of the anomalous steepness can then be attributed to this effect. Three years later, in 1958, Mills, Slee and Hill used their catalogue of sources to derive a new slope for the $\log N$–$\log S$

curve and they obtained the value that is largely accepted to-day, namely, -1.8 (although, in view of their errors, they regarded their results as compatible with a slope of -1.5). A slope of -1.8 is still anomalously steep, but it has since been confirmed by many further surveys, especially those made in Cambridge during the period 1959–68.

At the present time the most detailed analysis, and the one extending over the largest range of flux densities, is that due to Ryle and G. G. Pooley who used mainly the 5C survey to construct the $\log N$–$\log S$ relation shown in fig. 33. Their results show that the steep slope of -1.8 begins to flatten off at 4 flux units (at 408 MHz), and by 0.01 flux units has flattened off very greatly. This steepening and subsequent flattening have great significance for cosmology as we shall now see.

INTERPRETATION OF THE COUNTS

(*a*) *The three-halves power law for static sources.* This law depends on the sources being distributed uniformly. Suppose that there are ρ sources per unit volume and that they have an absolute radio luminosity per unit solid angle P at the frequency of observation. Then the sources whose measured flux density exceeds S comprise all the sources within a sphere of radius $(P/S)^{\frac{1}{2}}$ (because of the inverse square law). Their total number is then $\frac{4}{3}\pi\rho P^{\frac{3}{2}}S^{-\frac{3}{2}}$, and so the number N per unit solid angle is $\frac{1}{3}\rho\, P^{\frac{3}{2}}S^{-\frac{3}{2}}$. In practice we know that the sources have a large spread of absolute radio luminosity. This, however, does not affect the three-halves power law, which simply becomes

$$N = \tfrac{1}{3}(\Sigma\rho P^{\frac{3}{2}})\, S^{-\frac{3}{2}},$$

where the sum Σ is taken over the various luminosity classes.

(*b*) *The direct effects of the red shift.* There are three direct effects, all of which are of progressively increasing importance as S decreases.

(i) The effective value of P will depend on the red shift since we are observing in one small frequency-band radiation emitted in a different small frequency-band. Allowance must therefore be made for the spectrum of each source.

(ii) The red shift reduces the apparent brightness of a source over and above the effect of the inverse square law, so that the sphere corresponding to S is reduced in size. This has the effect of reducing N.

(iii) If the red shift is taken to imply an evolutionary universe with no creation of matter, then the sources were more congested in the past than they are now, that is, ρ was greater in the past. Now the greater the red shift the further the source, and so the longer we are looking into the past. Thus the effective ρ should increase with increasing red shift and so with decreasing S. This has the effect of increasing N.

In practice effect (i) is small. Most of the sources concerned have a spectrum of the form $S \propto \nu^{-0.7}$ (where ν is the frequency of observation) in the vicinity of the Cambridge observing frequency of 408 MHz. When we allow for the effect of red shift on the bandwidth we find that the effective luminosity P increases with red shift z like $(1+z)^{-0.3}$, a fairly weak dependence. By contrast, effects (ii) and (iii) are very important, and calculation shows that in all the cosmological models considered in this book (ii) is more important than (iii) (which is of course completely absent in the steady state model). Thus in all our cosmological models the direct effect of the red shift is to *flatten* the slope of the $\log N$–$\log S$ relation. In this way we arrive at a complete contradiction with the observations at large values of S.

One's first impulse might be to drop the assumption that the sources are distributed uniformly. One is free to do this in the sense that most of the sources concerned have not yet been identified optically, so that their red shifts and distances are unknown and one has therefore no direct observational knowledge of their distribution in space. However, there is a serious difficulty in doing this which arises from the fact that the source counts are fairly isotropic. It seems unlikely that irregularities could be present at such a scale and with such a disposition as to account for the steep slope of the $\log N$–$\log S$ curve and yet to preserve isotropy – unless one irregularity itself contains most of the sources. In that case approximate isotropy could be preserved if the edge of the irregularity in any direction occurs at such large red shifts that it is too faint to observe. We shall return to this exceedingly interesting question later in connexion with the distribution of QSOs and the isotropy of the cosmic microwave radiation. For the moment we shall neglect this possibility.

(*c*) *Evolutionary effects.* We must remember that in an evolutionary universe distant objects with large red shifts are being observed at an earlier stage in the development of the Universe than are nearby objects of small red shift. The possibility then arises that there has been a significant evolution in the intrinsic properties and distribution of radio sources in the time interval involved in the observations. Since we lack a detailed understanding of the origin and development of these sources we are free at this stage to suppose that the sources have evolved in whatever manner is required to account for the $\log N$–$\log S$ relation. In particular we would obtain a slope steeper than -1.5 if we assumed that in the past sources had on the average a sufficiently higher absolute luminosity P, or a sufficiently higher concentration ρ (over and above the kinematical effects of the expansion) than they have to-day. Such an explanation is clearly not available to the steady state theory, which requires P and ρ to have the same average values at all times and in all places in the Universe.

This point of view is closely similar to the one mentioned earlier in which the whole region out to some large red shift is simply part of one irregularity. Indeed the only difference concerns not the radio source counts but the nature of the Universe beyond the region surveyed in the counts. On the evolutionary view we see the same history of the Universe as any other observer, on the irregularity view we do not.

Various attempts have been made to construct evolutionary models that would reproduce the observed source counts. It has not been possible to find a definite model because the problem contains too many unknowns. On the cosmological side we do not know which model of the Universe is the correct one (see chapter 8). On the radio source side we have to take into account the very large spread in the absolute radio luminosity of the sources and the unknown rate of evolution which may itself differ for different classes of source. It may be different for radio galaxies and for QSOs, and it may be different for strong sources and for weak sources. Despite the uncertainties one can make the following points:

(i) No evolutionary model for the sources would be acceptable if it led to a greater *integrated* radio background than the observed diffuse extragalactic background. This is a severe constraint on possible models, as can be seen from the fact that the individual sources in the Ryle–Pooley survey already account for about half of the known extragalactic background at 408 MHz. The sources too faint to be detected in that survey may well account for most of the remaining half, since this would be consistent with a simple extrapolation of fig. 33.

(ii) The flattening-off in the slope of the $\log N$–$\log S$ relation at the faint end is presumably an indication that many of the faintest sources detected are at such large red shifts that effect (ii) (the weakening of the sources due to red shift) is beginning to dominate over evolutionary effects. How large these red shifts are is unknown, and different evolutionary models make different predictions. Likely values range between 3 and 5.

(iii) Most of the sources involved in the counts have not yet been optically identified. It follows that we cannot yet be sure how to compare the evolution of radio galaxies and QSOs. There is some preliminary evidence that the $\log N$–$\log S$ relation for QSOs has a steep slope, but this evidence is rather uncertain at the moment, and it would be valuable to have independent evidence that these objects are in fact evolving. Such evidence now exists. We shall discuss it in the next chapter along with other aspects of the distribution of QSOs.

7 The distribution of quasi-stellar objects

INTRODUCTION

INTRODUCTION

We saw in the last chapter that the anomalously steep slope of the log N–log S relation for radio sources is likely to have great cosmological significance. We also saw that the sources are of two kinds, the radio galaxies and the QSOs, and that both kinds are mixed up together in the log N–log S curve of fig. 33. The question naturally arises: what is the log N–log S relation for the two kinds of source taken separately? This is the first question we shall consider in this chapter.

THE LOG N–LOG S RELATION FOR QSOs

It must be admitted right away that the present optical identifications do not give us a complete separation of radio sources, even the brightest, into radio galaxies and QSOs. Accordingly any attempt to construct a log N–log S relation for radio galaxies and QSOs separately must be regarded as tentative. Nevertheless the results so far obtained are of great interest and are also entirely reasonable. P. Véron and Longair found independently in 1966 that the slope for the identified radio galaxies is close to -1.5, while the identified QSOs have an anomalously steep slope. This result for the radio galaxies is entirely reasonable since most of the identified ones have relatively small red shifts and so would be expected to obey the three-halves power law fairly well. It appears that anyone who sets out to explain the anomalously steep slope for the total counts must do it, in the first place, in terms of the QSOs.

One important question still remains: what is the log N–log S relation for radio galaxies of large red shift? This question is related to the problem of the nature of the unidentified 3C sources with good radio positions. Many of them lie in unobscured fields that are apparently empty down to the limit of the Palomar Sky Survey. The $\log N$–$\log S$ relation for these sources also has a steep slope, and for this and other reasons Véron believes them to be QSOs. Bolton on the other hand has reasons for believing that they are radio galaxies. This issue is clearly bound up with the problem of the source counts, but for the time being it remains unsettled.

We must now try to understand why the $\log N$–$\log S$ relation for QSOs has a steep slope. This question cannot be divorced from the problem of the nature and location of QSOs. There appear to be two main possibilities:

(i) If the QSOs are all cosmological then the steep slope is due either to large scale irregularities in their distribution or to evolutionary effects. The former explanation is unlikely, as we have already seen, because of the isotropy of the counts. The latter is quite reasonable because most of the QSOs have large red shifts and would be expected to show evolutionary effects.

(ii) If all the QSOs are local and uniformly distributed and if their $\log N$–$\log S$ relation is steep, then their red shifts cannot be gravitational in origin. The reason is that the gravitational red shift simply changes their effective absolute radio luminosity P; as we have seen (p. 85) the three-halves power law is independent of P and so would be expected to hold in this case. The red shifts must then be Doppler shifts associated with a local explosion or explosions. The $\log N$–$\log S$ relation would thus give information about these explosions and would have nothing directly to do with cosmology.

The reasons for doubting the local hypothesis have been discussed in chapter 5, and we shall in fact accept the cosmological hypothesis and along with it the idea that the QSOs evolve. The $\log N$–$\log S$ relation is not by itself a very sensitive probe of these evolutionary effects because of the large scatter in the intrinsic properties of the QSOs (figs. 34 and 35). A more effective method is to restrict attention to QSOs of known red shift. Despite the reduction in the number of usable sources the extra information available for each source gives one greater control over the analysis. To this analysis we now turn.

THE NUMBER–RED SHIFT RELATION FOR QSOs

When we know the red shift of a QSO we also know its distance in each of the standard cosmological models described in the next chapter. For each model we can then ask the question: are the QSOs of known red shift distributed uniformly in space? If they

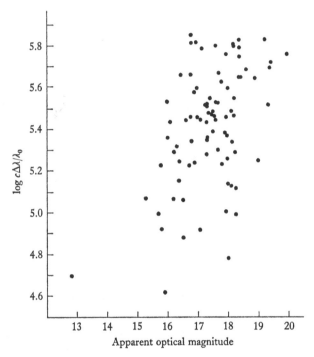

Fig. 34. The red shift–apparent optical magnitude relation for QSOs.

are not, and if we continue to discount the idea that we are situated inside a local irregularity, then we would have established that a systematic evolutionary process is occurring. This evolution would, of course, have to be compatible with that derived from the $\log N$–$\log S$ relation.

In performing analyses of this sort great care must be taken to avoid being misled by that bugbear of the observational astronomer, the intrusion of selection effects. In other words the data we have available to analyse may not be truly representative. Observers may consciously or unconsciously select for study QSOs with special properties that simplify their task or render it more interesting. For statistical discussions it is essential to have an unbiased sample of objects, or at least to be able to correct adequately for the selection effects that have in fact been introduced.

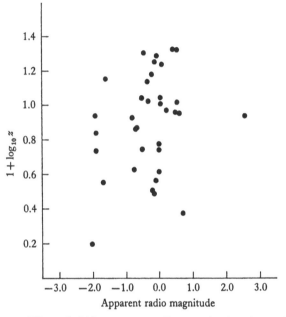

Fig. 35. The red shift–apparent radio magnitude relation for QSOs (m_R = 0.00 corresponds to $S = 10^{-25}$ wm^{-2} Hz^{-1}).

This problem is particularly severe in the analysis of the spatial distribution of QSOs because both radio and optical selection effects are present in the data. To minimise the radio selection effects we shall restrict ourselves to the 3C revised catalogue, which is believed to be complete down to 9 flux units at 178 MHz. Forty of these sources are QSOs with known red shifts. To minimise the optical selection effects we shall further restrict ourselves to those QSOs brighter than a visual magnitude V of 18.4. According to Schmidt the optical identifications of 3C revised sources are essentially complete down to this magnitude. This leaves us with a sample of 33 sources that we may take to be complete down to a known limiting radio and optical apparent magnitude.

Let us now assume a particular cosmological model. We can then compute for each of the 33 sources the maximum red shift at which it would still be included in our complete sample. In doing this, allowance must be made for the optical and radio spectra of each source, since we are observing at fixed optical and radio frequency bands, and therefore at different frequencies of emission, depending on the red shift. The maximum red shift for each source tells us the maximum volume V_m of space in which it could lie and still be included in our sample. If the sources are distributed uniformly they are equally likely to lie inside or outside the volume $\frac{1}{2}V_m$ around us. In this case the volumes V corresponding to the observed red shifts of each source should be such that the average value of V/V_m is $\frac{1}{2}$.

This test was carried out by Schmidt, who found that only 6 of the 33 sources have a V/V_m less than $\frac{1}{2}$, and that its average value is close to 0.7 for all the cosmological models of chapter 8. In other words there is an excess of QSOs at large red shifts as compared with a uniform distribution. Both the sense and the magnitude of the discrepancy are compatible with the provisional steep $\log N$–$\log S$ relation for QSOs.

The rate of evolution implied by Schmidt's results is very large. After allowing for the congesting effect associated with the expansion of the Universe Schmidt finds that the space–density of QSOs at a red shift of 1 is about one hundred times greater than the space-density in our locality. At a red shift of 1.5 the ratio lies between 300 and 800, depending on which is the correct model of the Universe. (The steady state model is, of course, ruled out by these results, if we are interpreting them correctly.)

Further evidence that there is an excess of QSOs at large red shifts has come from the very recent work of Braccesi. As explained in chapter 5 Braccesi has found that QSOs may be readily picked out from blue stars by their infra-red excess. He now has a substantial collection of objects selected in this way. Until their red shifts have been determined we can analyse them only in terms of their $\log N$–$\log S$ relation. In fact most of the sources are very faint or undetectable at radio wavelengths, so Braccesi has investigated the $\log N$–$\log S$ relation by taking for S their apparent *optical* luminosity. He finds the now familiar result that this relation has a steep slope, of order -1.8 (where radio units are used to express optical magnitudes). This implies that the space-density of QSOs evolves at a rate that is more or less independent of their radio properties.

This evolution in the properties of QSOs is clearly of great cosmological importance. It represents the first clear-cut direct evidence that the Universe in the past was different from what it is to-day. We would naturally like to go on to explain this particular evolutionary property. However, at the present time this is a completely open question. Despite some interesting suggestions that have been made, we have no understanding at all of the rapid change with time of the space-density of QSOs.

POSSIBLE LARGE-SCALE CLUSTERING OF QSOS

We now wish to discuss, in a very tentative manner, the possibility that QSOs occur in clusters just as galaxies do. Such clustering could not by itself explain the anomalous $\log N$–$\log S$ relation because of the isotropy problem, but it may exist nevertheless. The reason is that the isotropy studies refer to sources defined only in terms of a limiting flux density. Because of the large spread in absolute luminosity these sources include nearby faint ones and distant bright ones and so the observed isotropy relates to some average taken over a large distance in depth (and over radio galaxies and QSOs as well). The clustering might then be real, but be averaged out by this procedure. Clearly a more efficient way of testing for clustering would be to select sources in a fairly small red shift range, and to see whether they are distributed at random over the accessible part of the sky.

Such a procedure was in fact carried out at the end of 1966 by P. Strittmatter, J. Faulkner and M. Walmesley without the problem of clustering in mind. They pointed out that QSOs with large red shift did not appear to be randomly distributed. The QSOs with red shift $z > 1.5$ were mainly confined to two regions, one near the North Galactic Pole and the other in the South Galactic Hemisphere. These groups each had an angular diameter of about 30°. QSOs with intermediate z also appeared to be distributed anisotropically, but with a larger angular diameter. Finally, for QSOs with low z there was no appreciable anisotropy (see fig. 36). More

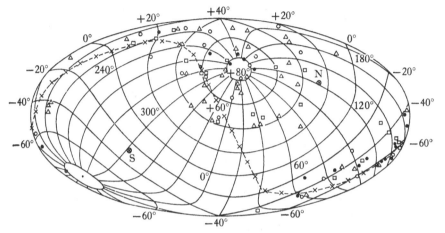

Fig. 36. The distribution of QSOs in the sky in galactic co-ordinates. N and S refer to the Earth's north and south poles, and the broken line represents the projection of the Earth's equator on the sky. ●, $z > 1.5$; ○, $1.5 > z > 1$; △, $1 > z > 0.5$; □, $z < 0.5$. (From *Quasi-Stellar Objects* by Geoffrey Burbidge and Margaret Burbidge. W. H. Freeman and Company. Copyright © 1967.)

recent red shift data have weakened this picture, but it has not altogether lost its suggestive power.

One's first thought is that the result is due to selection effects, and most astronomers probably believe that this is so. However, Strittmatter, Faulkner and Walmesley were able to show that the more obvious selection effects, at least, were not responsible. Since new red shift data are steadily being obtained and being added to the picture it seems better to wait to see whether the trend is maintained rather than search for some more complicated selection effect. The point of discussing the question here is simply that if the trend is maintained, then the consequences would be very interesting.

The discoverers of the trend appear to have been thinking in terms of two definite groups of QSOs in two definite directions, in other words, in terms of marked anisotropy in their distribution. They suggested that if the trend is confirmed by further observations, then either (a) the Universe is anisotropic for $z \sim 1$, or (b) QSOs are not at cosmological distances. In view of the high degree of isotropy of the cosmic microwave background (chapter 16) possibility (a) now seems unlikely. On the other hand possibility (b) with its suggestion of an explosion in two favoured directions would be a reasonable one on dynamical grounds, were it not for the other difficulties encountered by the local hypothesis.

Fortunately there exists an alternative interpretation for the trend if it is maintained, namely, in terms of inhomogeneity rather than anisotropy. It is possible that QSOs occur in clusters on a scale corresponding to a red shift of about unity, that is, to a distance of about 2000 Mpc. Since QSOs are on the average about a hundred megaparsecs apart the suggested scale of their clustering would be much the same in relation to their separation as is the case for the clustering or super-clustering of galaxies.

It would be natural to associate the clustering of QSOs (if such clustering is found) with the existence of large scale irregularities in the distribution of matter in the Universe, the space-density of QSOs acting as a probe for the density of matter. The question then arises: is it reasonable to contemplate density fluctuations in the Universe on a scale > 1000 Mpc? This is a difficult question to answer because we do not yet understand the origin of the smaller-scale fluctuations (galaxies and galaxy clusters) that we know to exist in the Universe. However, it may be helpful to consider what is known, or reasonably conjectured, about these smaller-scale fluctuations. This information is summarised, in very round figures, in table 3 (where we have assumed the present mean density of the Universe to be 10^{-30} g cm^{-3}).

It does not seem out of the question that on a scale ~ 1000 Mpc there should exist QSO clusters in which the density contrast is, say, two to one. Moreover what little insight has been gained into the way fluctuations develop as the Universe expands suggests that the largest fluctuations are less damped by radiative viscosity

TABLE 3   *Observed clustering in the Universe*

Scale	Object	Density contrast $(\delta\rho/\rho)$
30 kpc	Galaxy	10^6
1 Mpc	Galaxy cluster	10^3
30 Mpc	Supercluster	10
1000 Mpc	?	?

than the smaller ones. It thus seems quite sensible to analyse the observations for large scale fluctuations. However, we must await the discovery of many more QSOs, and the determination of their red shifts, before such an analysis could be profitably carried out.

A further reason for discussing this possibility is its interesting relation with the angular distribution of the cosmic microwave background (chapter 16). If the large scale clustering exists it would tend to produce anisotropies in this background. The reason is that if the background radiation comes to us through a cluster it would suffer a gravitational red shift, the potential as it enters a cluster being different from the potential as it leaves owing to the expansion of the cluster with the Universe. This red shift would be different in different directions since the arrangement of clusters would depend on direction. As explained in chapter 16 the effective temperature of the background is closely related to the red shift, and so this temperature would also depend on direction. The existing measurements on the angular distribution of the background temperature are not quite accurate enough to give us the information we need. There remains the enticing prospect that further observational studies of this type may give us information of a totally new character concerning the large scale distribution of matter in the Universe.

8 Models of the Universe

INTRODUCTION

We now face the formidable task of constructing a theoretical framework in which to set the observations of the expanding Universe. Technical difficulties are to be expected in dealing with such a complex mass of material, but what makes theoretical cosmology at once so exciting and so frustrating are the conceptual difficulties that arise. Some of these are obvious. Since we are studying a very large system, spanning thousands of megaparsecs in distance and billions of years in time, we have to ask the question: do the laws of physics, found to hold good here and now, have universal significance? Are they the same not merely a long way away and a long time ago but in a Universe which then had a very different appearance? Even if we accept as a working hypothesis that the laws are universal another question arises. Have we in fact determined the local laws with sufficient precision to enable us to apply them without appreciable error on a very large scale? This is the macroscopic counterpart to the question that arises in microscopic (atomic and nuclear) physics. The laws that apply to man-sized objects do not apply without modification on the atomic scale. On the other hand the atomic laws do apply to man-sized objects, but many of their characteristic features become nearly averaged out when enormous numbers of atoms are involved. In principle, however, one could imagine deducing the atomic laws from a sufficiently accurate and detailed study of man-sized objects.

The same problem arises when we investigate the Universe. Effects which may be very small on the scale of the laboratory could be dominant on the scale of the Universe. Thus we could add small terms to the basic equations of physics which would be negligible locally, but which could have a decisive effect on the solutions in the large. One example that we shall meet is the so-called cosmical constant. What chance have we of using the correct equations to describe the Universe without a very detailed prior knowledge of its behaviour? We are in a similar position to that of the physicists at the turn of the century, who tried to describe atomic behaviour entirely in terms of Newtonian dynamics.

As if this were not enough there are subtler difficulties that arise when we go over from considering the Universe simply as a very large system to considering it as the totality of what exists. If this totality is a legitimate concept it introduces two differences from our usual way of doing physics. First of all the system we are studying is not immersed in an environment whose properties are regarded, for any given problem, as fixed by outside considerations. For instance, if we are studying star formation we accept that interstellar clouds, or at least the Galaxy, exist, and that their properties help to determine the properties of stars. By contrast, the Universe as a whole is self-contained, and all its properties must, as it were, be accounted for in terms of itself.

There is a second way of arriving at the same conclusion, namely, from the fact that the Universe as a whole is unique. We may usefully contrast its behaviour with that of a projectile at the surface of the Earth. We know that such a projectile can start out in any direction and with a large range of possible velocities. The laws governing projectiles must then be flexible enough to permit essentially all directions and velocities to occur in practice. No actual trajectory is of fundamental significance, only the properties common to all trajectories (such as being parabolic). These common properties are then enshrined in the laws of physics. In mathematical terms the law is expressed by a differential equation and the individuality of each case by the appropriate choice of initial or boundary conditions.

By contrast, we do not have available many universes, some expanding, some contracting, some more non-uniform than others, and so on, whose common properties can be established by observation and then enshrined in laws of nature. We have only one

Universe, so the significant fact in this case is the actual behaviour of this single phenomenon. The laws of nature must thus be formulated in such a way that they relate only to the actual Universe, for other universes by definition cannot exist. In other words, we should seek a theory which describes all that actually happens and nothing that does not, a theory in which everything that is not forbidden is compulsory.

Such a rigid theory has not yet been discovered. For instance, general relativity, which is the best theory of space, time and gravitation that has so far been proposed, is, as we shall see, consistent with an infinite number of different possibilities, or models, for the history of the Universe. Needless to say, not more than one of these models can be correct, so that the theory permits possibilities that are not realised in Nature. In other words, it is too wide. We can put this in another way. In the absence of a theory anything can happen. If we introduce a weak theory too many things can still happen. A strong enough theory has not yet been discovered.

There is one other problem which has excited much interest, and that is the possibility of a direct connexion between elementary particle physics and cosmology. Apart from general reasons for perhaps expecting such a connexion to exist there is one specific reason, namely, the famous numerical coincidences between atomic and cosmological quantities. These coincidences are discussed at the end of this chapter; here we merely mention that they involve the charge and mass of an electron, the velocity of light, the gravitational constant and the Hubble constant.

Many approaches to these profound problems have been made in the last 40 years. Although none of them has been very successful they are of great intellectual interest. Nevertheless they will not be described in this book. Instead we shall take an intensely pragmatic view, and simply assume that the locally established laws hold good everywhere and at all times without any modification. We shall accept the difficulties that go with this view. For instance we shall try to determine the initial conditions that prevailed when the Universe was very dense without attempting to understand why those were the conditions that did prevail. We shall also try to understand the numerical coincidences as far as possible in terms of known physics. There is a very good reason for taking such a pragmatic view at the present time. As we stressed in the preface to this book, cosmologists are now faced with a great flood of new observations which have to be accommodated into their system. The first question must be: can these observations be understood in terms of the known laws of physics? While this question is in the process of being answered, and while indeed new observations are still coming in, it is not the time to be asking great questions of principle. Let us hope that when these observations have been digested, and we look at the Universe from a new point of vantage, these questions of principle can be answered.

THE NEWTONIAN DYNAMICS OF A LARGE GAS CLOUD

To prepare the way towards understanding relativistic cosmology we shall consider the Newtonian dynamics of a large gas cloud. Not only is the Newtonian theory mathematically simpler, it also leads to many results that are essentially the same as in relativity, as was discovered in 1934 by E. A. Milne and W. H. McCrea. The relation between the gas cloud on the one hand and the galaxies and QSOs that make up the Universe on the other, need not be specified too closely. One can think of the galaxies and QSOs either as the particles of the gas, or as being localised condensations in an actual intergalactic (atomic or ionised) gas, condensations that act as tracers for the average motion and perhaps the average density of the gas in their general vicinity.

What is important is that the gas cloud should not be taken to be infinitely large. As Newton discovered, his dynamics and gravitational theory run into difficulties when applied to an infinite system. For instance the gravitational potential at a point due to all the matter in the system would be infinite. This difficulty does not arise in general relativity, but here we may avoid it by taking our cloud to be large but finite in size. Another way of avoiding it was proposed in the last century by C. Neumann and H. Seeliger, who added to the Newtonian gravitational force a repulsive force directly proportional to the distance of a particle from the origin and independent of the physical properties of matter. In accordance with our philosophy of not tampering with the laws of physics we shall not adopt this device here.

There is one important difference between a large cloud and an infinite one. The large cloud has a unique centre while the infinite one does not. We do not really want a special point to be picked out in this way, and we can minimise the effect of having one by making the cloud uniform out to its edge, isotropic about its centre and much larger than any distance that has yet been measured. Under these circumstances any galaxy or QSO that we can detect would see around itself with arbitrarily high precision a uniform isotropic Universe. In the last analysis we can never distinguish observationally between an infinite Universe and a finite one that is suitably larger than any distance yet surveyed. The debatable assumption that we have made is then not so much that the system is finite as that it is uniform and isotropic. For the moment we may regard these assumptions as being useful for a first attack on the theoretical problem. The extent of their agreement with observation and the reasons for this agreement will be discussed later.

The assumptions of uniformity and isotropy are useful not only for permitting any point in the observable region of the cloud to regard itself as the centre. They also very much simplify the motion of the cloud as seen by any observer moving with the cloud. In fact the velocity v of a particle at a radius vector r from a comoving observer then satisfies the simple relation

$$v = f(t)r, \tag{1}$$

where f is any function of the time t. Thus *at a given time* the motion of the particles with respect to any one co-moving particle satisfies a linear velocity–distance relation, a relation that we have already called the Hubble law. This result shows how stringent are the assumptions of homogeneity and isotropy. Of course we can regard this linear law as a first-order approximation to a more complicated law, but it does seem to be a reasonable approximation to what is observed. In fact when we test it out to distances large enough for a term in r^2 to perhaps become appreciable we must also allow for the fact that we may be looking so far back in time that the quantity $f(t)$ can no longer be taken to be constant. Thus we could absorb a small non-linear term by slightly modifying f.

It is useful for our further development to integrate the velocity law (1) to give the position of the particle at time t. The result is

$$r = R(t)r_0, \tag{2}$$

where $R(t)$ is related to $f(t)$ by the equation

$$\frac{1}{R}\frac{dR}{dt} = f(t), \tag{3}$$

as is easily seen by differentiating (2) and comparing it with (1). In (2) r_0 is the position of the particle at some standard time t_0, and so

$$R(t_0) = 1.$$

We see from (2) that the only possible motions consistent with uniformity and isotropy are those of uniform expansion or contraction, a simple scaling up or down with a time-dependent scale factor $R(t)$. To simplify the notation we shall write (3) in the form

$$\frac{R}{R} = f(t)$$
$$= \frac{1}{\tau(t)},$$

so that

$$v = r/\tau. \tag{4}$$

Notice that the Hubble constant τ is independent of r (that is what is meant by the Hubble law) but does depend on t.

So far we have managed without using Newton's second law of motion or his law of gravitation. Everything has followed from our symmetry assumptions alone, and we may regard our results up to this point as kinematical in character. They show that the whole motion of the cloud is determined by just one arbitrary function of the time. To determine this function more closely we must introduce dynamical considerations. This problem is much simplified by the well-known Newtonian result that in a uniform isotropic system the gravitational force acting on a particle at the position r relative to the centre is entirely due to the matter lying closer to the centre than does the particle. The effect of the matter outside this sphere cancels out by symmetry. We can use this fact to obtain a simple result for the dynamics of the cloud as seen from its centre, and we can transfer this result to any other origin within the observable region by using the relation (4). In this way we avoid the tricky question of whether all the co-moving observers, who will be in acceleration relative to one another, represent inertial frames of reference.

Newton's laws lead to the following equation for the scale-factor $R(t)$:

$$R^2 \ddot{R} + \frac{4\pi}{3} G\rho(t_0) = 0, \qquad (5)$$

where G is the Newtonian constant of gravitation and $\rho(t_0)$ is the density of the cloud at the standard time t_0 which, because of the conservation of matter, satisfies the equation

$$\rho(t) = \frac{\rho(t_0)}{R^3(t)}, \qquad (6)$$

since $R(t_0) = 1$. Our dynamical equation (5) for R shows immediately, what is intuitively obvious, that we cannot have a static cloud ($\dot{R} = \ddot{R} = 0$) unless $\rho = 0$. We are thus led to expect a systematic motion of expansion or contraction on a scale over which the Universe is approximately uniform. In fact this result, translated into relativistic terms, was derived rigorously from general relativity before the Hubble law was established observationally. It might seem at first sight that a star or a planet, which are at least quasi-static, constitute counter-examples. However, in these cases gravity is balanced by a pressure gradient, which cannot exist in a uniform system. In any case the enormous pressure gradient needed to stabilise the Universe is obviously not in fact present.*

Fortunately the equation for $\ddot{R}$ can be integrated easily to give a dynamical equation for $\dot{R}$, the rate of expansion or contraction. To effect this integration we multiply (5) by $\dot{R}/R^2$ and so obtain the integrated equation

$$\dot{R}^2 = \frac{8\pi}{3} \frac{G\rho(t_0)}{R} - k. \qquad (7)$$

* Moreover in relativity the pressure contributes (positively) to the gravitational field. In an ordinary star this contribution is negligible, but a pressure gradient which at first sight might stabilise the Universe would actually increase the effective gravity.

Here k is a constant of integration which is a measure of the total energy (kinetic plus potential) of a particle. With the sign we have chosen for k (which is the conventional one) the cloud is gravitationally bound or unbound according as k is positive or negative. When k is zero the kinetic and potential energies are equal and opposite, and the cloud can just expand to infinity. We now consider these three cases in more detail. This is worth doing *because relativity leads to the same equation for the scale-factor $R(t)$*, although the constant k then has a somewhat different meaning.

(i) $k = 0$. In this case

$$\dot{R}^2 = \frac{8\pi}{3} \frac{G\rho(t_0)}{R},$$

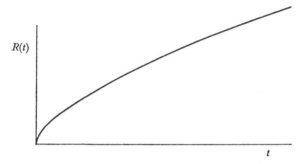

Fig. 37. The scale factor of the Universe $R(t)$ for a Newtonian model with zero total energy, or a relativistic model with zero pressure and space curvature (the Einstein–de Sitter model).

and $\dot{R}$ tends to zero as R tends to infinity. This equation for $\dot{R}$ can be integrated to give the explicit time-dependence of R. We find

$$R \propto t^{\frac{2}{3}},$$

where the constant of proportionality is $\{6\pi G\rho(t_0)\}^{\frac{1}{3}}$. The graph of $R(t)$ is shown in fig. 37. In relativity this relation characterises the well-known Einstein–de Sitter model of the Universe.

(ii) $k > 0$. In this case the cloud is gravitationally bound and expands out to a maximum size R_{max} which occurs when $\dot{R} = 0$, that is, when

$$R_{\text{max}} = \frac{8\pi}{3} \frac{G\rho(t_0)}{k},$$

at which point the motion reverses into a collapse. This collapsing phase of the motion is familiar to astronomers concerned with star formation, since they are interested in a finite cloud collapsing from rest. The solution of (7) is fairly simple and one finds that the $R(t)$ curve is a cycloid, as in fig. 38. This gives us an oscillating Universe, although there is no justification in adding further cycles to the oscillation as some writers do. We shall return later to the difficult question of handling the singular point $R(t) = 0$ correctly.

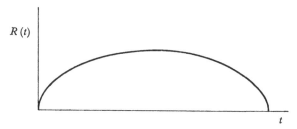

Fig. 38. The scale factor of the Universe for a gravitationally bound model.

(iii) $k < 0$. In this case

$$\dot{R}^2 = \frac{8\pi}{3} \frac{G\rho(t_0)}{R} + (-k) \quad (-k \text{ a positive constant}). \quad (8)$$

Hence as R tends to infinity the rate of expansion $\dot{R}$ tends to a non-zero positive quantity. In other words, the particles have excess kinetic energy and are still moving apart when the cloud is infinitely large and dilute. Unfortunately (8) cannot be integrated in simple terms, but R does depend simply on t for small t and large t. For small t, R is small and the first term on the right hand side of (8) dominates over the second term. This gives us the same equation as in case (i) with $k = 0$, and we have

$$R(t) \propto t^{\frac{2}{3}} \quad (t \text{ small}).$$

For large t, R is large and the second term dominates. The equation then integrates to give

$$R \propto t \quad (t \text{ large}),$$

which corresponds to unaccelerated expansion, gravitation being negligible, for large t. A sketch of $R(t)$ over the whole range of t is given in fig. 39.

We now consider some of the observable properties of these models in order to see which fits the actual Universe best.

(a) *Expansion rate.* We have already seen in (4) that the Hubble constant τ is given by $\quad \tau = R/\dot{R}.$

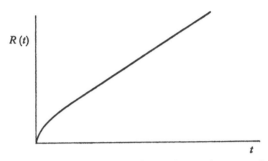

Fig. 39. The scale factor of the Universe for a gravitationally unbound model.

If we choose the present time as the standard time t_0, then the present value of R is unity, and the present value of τ is $1/\dot{R}$. We thus know that the slope of the $R(t)$ curve at present is about $(10^{10}\text{ years})^{-1}$.

(b) *The present age of the Universe.* Fig. 40 shows geometrically that the time in the past when R was zero, which we may call the age of the Universe, is less than τ in all the models. This is obvious physically since the Universe was expanding faster in the past because of its self-gravitation (see (c) below). In the Einstein–de Sitter model ($k = 0$) we have $R \propto t^{\frac{2}{3}}$, and so

$$\tau = \frac{R}{\dot{R}} = \tfrac{3}{2}t.$$

With $\tau \sim 10^{10}$ years, we have $t \sim 6.7 \times 10^9$ years, which is somewhat less than the ages of the oldest stars in the Galaxy, and dangerously close to the age of the Sun. This time-scale difficulty would be relieved if the actual value of the Hubble constant were somewhat larger than 10^{10} years. In view of the great difficulty in determining this quantity such a revision is by no means ruled out.

The time-scale difficulty is even worse in the oscillating models ($k > 0$), where $t < 2/3\tau$. By contrast in the ever-expanding models ($k < 0$), t approaches τ more nearly as the present density $\rho(t_0)$ is

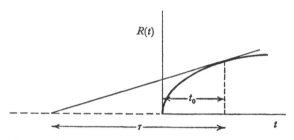

Fig. 40. The age of the Universe t_0 is always less than the Hubble constant τ (which is determined by the tangent to the $R(t)$ curve).

lowered, the effects of gravity are correspondingly reduced, and the approximation $R \propto t$ is valid from an earlier starting time. If future work confirms the present value of the Hubble constant this is the model that would be favoured.

(c) *Deceleration parameter.* In all our models the expansion of the Universe is decelerating because of self-gravitation. The amount of deceleration gives us a useful measure of the amount of self-gravitation and so of the material density. The deceleration is essentially $-\ddot{R}$, but it is convenient to define it in such a way that it is independent of the time t at which we set $R(t_0) = 1$, and also such that it is dimensionless. We achieve the first aim by considering the quantity $-\ddot{R}/R$. This has dimensions (time)$^{-2}$, so to obtain a dimensionless quantity we multiply by $R^2/\dot{R}^2\,(=\tau^2)$. We thus take as our definition of the deceleration parameter q the equation

$$q = -\frac{R\ddot{R}}{\dot{R}^2}.$$

This is a purely kinematical definition. If we adopt the dynamical laws (5) and (6) and use (4) we find

$$q = \frac{4\pi}{3} G\rho\tau^2, \tag{9}$$

which shows the way in which q measures the density of matter ρ. In general q changes with time but an interesting exception is the Einstein–de Sitter model ($k = 0$) in which $R \propto t^{\frac{2}{3}}$, and so $q = \frac{1}{2}$ at all times. We have in fact the following relations

model (i)	($k = 0$)	$q = \frac{1}{2}$,
model (ii)	($k > 0$)	$q > \frac{1}{2}$,
model (iii)	($k < 0$)	$0 < q < \frac{1}{2}$.

(*d*) *Density*. This is closely related to the deceleration parameter, as we have just seen.

Model (i) ($k = 0$). We have $q = \frac{1}{2}$ and so from (9)

$$\frac{8\pi}{3} G\rho\tau^2 = 1.$$

Thus this model has the important property of leading to a definite value for the present density once the present value of the Hubble constant τ is known. Taking $\tau = 10^{10}$ years we have

$$\rho = 2 \times 10^{-29} \text{ g cm}^{-3},$$

a value that has figured in many astrophysical investigations of the contents of the Universe. It takes on special interest because the smoothed out density of the known matter in galaxies ρ_g is believed to be rather less than this. In fact

$$\rho_g \leqslant \sim 10^{-30} \text{ g cm}^{-3},$$

according to present-day estimates (chapter 3). This discrepancy has led to many speculations about the nature of the 'missing matter' (if it exists), and especially about the possibility that intergalactic space contains a gas density $\sim 2 \times 10^{-29}$ g cm^{-3}. These questions will be discussed in chapters 9 and 10.

Model (ii) ($k > 0$). In this type of model gravity is dominant and the present density exceeds the critical value of 2×10^{-29} g cm^{-3}, thus accentuating the problem of the 'missing matter'. It also suffers most from the time-scale difficulty, as we have seen.

Model (iii) ($k < 0$). In this type of model gravity is more or less unimportant except in the earliest stages and the present density can have any value less than the critical one of 2×10^{-29} g cm^{-3}. Of course observations of galaxies give a lower limit of ρ_g for the density, and if this is close to the actual density we would have for the present value of the deceleration parameter $q_0 < \frac{1}{40}$, which is small compared with the value of $\frac{1}{2}$ expected on the Einstein–de Sitter model. Accordingly the Universe would now be expanding at a rate which would remain nearly constant in the future. Such a model has the advantage that its present age is close to τ, and so is nearest to solving the time-scale difficulty.

This completes our discussion of the Newtonian dynamics of a large gas cloud. The reader will have noticed one very important omission. We have not discussed the behaviour of light in these models, and have therefore said nothing about the red shift, the apparent brightness of a distant source, and so on – the very properties that most closely link together observation and theory. The reason for this omission is that Newtonian theory does not provide us with a satisfactory account of the behaviour of light. Although the relativistic formula for the red shift is a very simple one, attempts to derive it in a Newtonian setting give one the uneasy feeling that one is stretching the Newtonian concepts too far. As a result one has no faith at all in the answer until it is checked by relativity. Under these circumstances there seems to be no point in studying further our over-simplified picture. We must therefore now turn our attention to relativistic cosmology. The discussion necessarily becomes somewhat more mathematical, and if the reader is uninterested in following the details he is advised to turn to p. 117 which gives a summary of the results we obtain.

RELATIVISTIC COSMOLOGY

General relativity* differs from the Newtonian theory of gravitation in the following respects:

* For a more detailed account see my book *The Physical Foundations of General Relativity*.

(i) It is based on ten potentials instead of one (or Maxwell's four in electrodynamics).

(ii) It is a non-linear theory, the total effect of several bodies not being the simple sum of their separate effects.

(iii) Pressure as well as density is a source of gravitation.

(iv) It is usually expressed in geometrical language, the ten potentials giving the metrical properties of space–time, which is curved in the presence of a gravitational field.

(v) There is no difficulty in having a gas cloud fill the whole of space.

Some of these differences can lead to great mathematical difficulties, but fortunately these are minimised when we impose the powerful symmetry assumptions of uniformity and isotropy as we did in the Newtonian discussion. These symmetry assumptions alone limit the metric (which gives the four-dimensional distance between neighbouring points of space–time) to the following form:

$$ds^2 = c^2\,dt^2 - \frac{R^2(t)}{(1+\tfrac{1}{4}kr^2)^2}\{dr^2 + r^2(d\theta^2 + \sin^2\theta\,d\phi^2)\},$$

as was shown by H. P. Robertson and A. G. Walker following the pioneering work of E. A. Milne. This differs from the special relativity metric for Minkowski space–time only by the presence of the undetermined scale factor $R(t)$ and the constant k. It is clear in a general way that the scale-factor $R(t)$ has much the same meaning here as in the Newtonian theory. To see this consider the Universe at a particular time t_0. Then we have $dt = 0$, and the metric for 3-dimensional space at the time t_0 becomes

$$ds^2 = -\frac{R^2(t_0)}{(1+\tfrac{1}{4}kr^2)^2}\{dr^2 + r^2(d\theta^2 + \sin^2\theta\,d\phi^2)\}.$$

At a later time t_1 we would have exactly the same metric, except that every interval ds would be multiplied by the factor $R(t_1)/R(t_0)$. If this factor is greater than 1, intervals are increasing with time and we have an expanding Universe.

The meaning of the quantity k is rather different here, however. Gravitational potential energy is an elusive concept in general relativity, and it is best to think of k as giving the *curvature* of

3-dimensional space at any time t_0. We then have the following possibilities:

(i) $k = 0$. Three-dimensional space is then Euclidean. In particular the surface area of a sphere of radius r is $4\pi r^2$.

(ii) $k > 0$. In this case the geometry of space is said to be spherical. It is in fact the 3-dimensional analogue of the geometry on the surface of a sphere. On such a surface a circle is the locus of

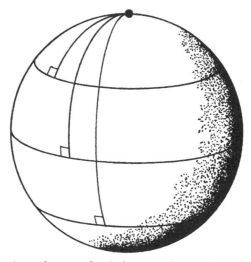

Fig. 41. The circumference of a circle on a sphere is less than 2π times its radius (as measured along part of the great circle). The corresponding geometry is non-Euclidean, and has positive curvature.

points at a constant distance from a given point, distance being measured along great circles (fig. 41). The circumference of such a circle is less than 2π times its radius. The difference is small when the radius is small, but for a larger radius it becomes substantial until the radius goes one-quarter way around the sphere, when the circumference is a maximum. For a larger radius still the circumference gets smaller again, and goes to zero as the radius goes half way round the sphere (fig. 41). In a similar way in the 3-dimensional spherical space the surface area of a sphere of radius r is less than $4\pi r^2$, grows to a maximum as r increases and then shrinks to zero again. The volume of the space is finite and is in fact equal to $\pi^2 R^3$, so it increases as R increases with time.

(iii) $k < $ o. In this case the geometry of space is said to be hyperbolic. The surface area of a sphere of radius r is greater than $4\pi r^2$. The volume of the space is infinite except in pathological cases that need not concern us.

Our results so far have been entirely kinematical. They place no restriction on R as a function of time, nor do they impose any relation between $R(t)$ and k. To proceed any further we must use the relativistic analogue of Newton's law of gravitation, that is, we must use Einstein's field equations. At one time, before the expansion of the Universe had been discovered, Einstein proposed a modification of his field equations that would permit the Universe to be static ($R = $ constant). The extra term involved an undetermined constant (the cosmical constant) and for a suitable choice of its sign its effect would be to oppose self-gravitation and so permit a static solution. The value of the constant required to achieve this situation was so small that its presence in the field equations would not disturb the agreement between general relativity and observation in the solar system. This is an example of the difficulty we mentioned earlier, that the local laws can be modified in a way that does not disturb their agreement with observation, but can be of decisive significance on the cosmological scale. In accordance with our philosophy in this book we shall not introduce a cosmical constant into the field equations.*

In view of the fact that in relativity pressure acts as a source of gravitation we must be careful now to specify the pressure in the gas cloud that represents the matter in the Universe. This pressure can be taken to include contributions from the peculiar motions of the galaxies, from the intergalactic gas (which may be hot), from radiation, and from intergalactic magnetic fields and cosmic rays. At the present time these pressures are almost certainly unimportant as a source of gravitation in comparison with the energy-density of the matter in galaxies. However, for the moment we shall

---

* However, one of the possible models of the Universe when the cosmical constant is kept – the Lemaître model – has recently attracted considerable attention in connexion with the mysterious absorption red shift of 1.95 in the spectra of QSOs (p. 77).

keep the pressure p in as an unknown. Einstein's field equations now lead to the following relations:

$$\dot{R}^2 = \frac{8\pi}{3} G\rho R^2 - k, \tag{10}$$

$$2\frac{\ddot{R}}{R} + \frac{\dot{R}^2}{R^2} = -8\pi \frac{Gp}{c^2} - \frac{k}{R^2}. \tag{11}$$

The first of these equations resembles our Newtonian equation (7) and would indeed be identical to it if we could write, as before,

$$\rho(t) = \rho(t_0)/R^3.$$

However, we must remember that if the pressure does work during the expansion, this work will change the energy density and so, in relativity, it will change the matter density ρ. At this point then we set $p = 0$ for simplicity. If we now multiply (11) by $R^2\dot{R}$ we get

$$2R\dot{R}\ddot{R} + \dot{R}^3 = -k\dot{R},$$

which integrates immediately to

$$R\dot{R}^2 = -kR + \text{constant}.$$

Comparison with (10) now shows that

$$\rho R^2 = \frac{\text{constant}}{R},$$

so that $\rho \propto 1/R^3$, as we want. Thus in the case of zero pressure the governing equation for the scale-factor R is

$$\dot{R}^2 = \frac{C}{R} - k,$$

where $$C = \frac{8\pi}{3} G\rho R^3 = \text{constant}.$$

Hence despite all the differences between general relativity and Newtonian theory, *the scale-factor R satisfies the same equation in both theories, so long as the pressure is negligible.* This is the Milne–McCrea theorem. Accordingly, the classification of models and the time-dependence of R is the same in both theories, and we need not repeat our discussion of these questions. Since in the present case the gas cloud fills the whole Universe at all times, it is better not to speak of bound and unbound clouds for $k > 0$ and $k < 0$, but rather of spherical and hyperbolic space, or closed and open space, or oscillating and ever-expanding space.

Two further models that are not pressure-free deserve mention at this point. One concerns the important physical case in which radiation dominates completely over matter as a source of gravitation. This may well be the situation that existed in the early stages of our own Universe, a question that is discussed in chapter 12. The pressure would now no longer be negligible, and in fact $p/c^2 = \frac{1}{3}\rho$. We can eliminate ρ by adding (10) and (11) to obtain

$$\frac{\ddot{R}}{R} + \frac{\dot{R}^2}{R^2} + \frac{k}{R^2} = 0.$$

As before we can neglect k for sufficiently small R, and then we can integrate the equation to obtain

$$R \propto t^{\frac{1}{2}} \quad (t \text{ small}).$$

This corresponds to a more rapid expansion than when only pressure-free matter is present ($R \propto t^{\frac{2}{3}}$ (t small)), because the pressure of radiation exerts its own gravitational field, thereby increasing the amount of gravity acting. This increases the rate of expansion, as is obvious if we reverse the sense of time and consider the resulting rate of collapse.

The second model with pressure that we wish to mention has $p/c^2 = -\rho$, that is, it contains a tension rather than a pressure. The corresponding gravitational effect is now repulsive, and in fact we obtain a model whose expansion is accelerating rather than decelerating. We shall in addition take $k = 0$ to obtain the model we want. We can then eliminate ρ from (10) and (11) to obtain

$$\frac{\ddot{R}}{R} - \frac{\dot{R}^2}{R^2} = 0,$$

or $$(\ln \ddot{R}) = 0.$$

Hence $$\ln R = t/\tau + b \quad (\tau, b \text{ constants}),$$

and $$R \propto e^{t/\tau}.$$

This model differs from our previous ones in that R does not go to zero a finite time ago (fig. 42). It is the celebrated de Sitter model (not to be confused with the Einstein–de Sitter model which has $R \propto t^{\frac{2}{3}}$, $k = 0$). The exponential curve is self-similar, that is, one cannot tell where one is along it by intrinsic measurements; it has no natural origin. That is why the de Sitter metric forms the basis of the steady state theory of H. Bondi, T. Gold and F. Hoyle.

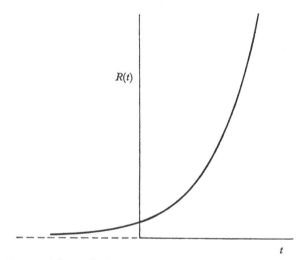

Fig. 42. The scale factor of the Universe in the de Sitter (steady state) model.

In this theory the Universe does not evolve from a dense state to a dilute one. The reason is that (10) with $k = 0$ would give

$$\frac{8\pi}{3} G\rho\tau^2 = 1.$$

Although this is the same relation as in the Einstein–de Sitter model, there is an important difference, for here τ is independent of time and so ρ is also independent of time, whereas in the Einstein–de Sitter model $\tau \propto t$ and $\rho \propto 1/t^2$. In the steady state model the density ρ remains constant because the work done by the tension during the expansion results in a continuous 'creation' of matter. This just compensates for the diluting effect of the expansion.

The steady state model is attractive in many ways, although the physical origin of the tension has never been satisfactorily explained. However, the recent evidence from the radio source counts (chapter 6), the red shifts of the QSOs (chapter 7) and the cosmic microwave radiation (chapter 14) all tell heavily against it, and we shall consider it no further.

SUMMARY OF NEWTONIAN AND RELATIVISTIC COSMOLOGY

Our discussion of Newtonian and relativistic cosmology has been rather elaborate, and many readers will probably be content with a brief summary that would suffice to make the rest of this book intelligible. We therefore provide such a summary here before going on to consider the propagation of light in the various models.

We assume the Universe to be uniform and isotropic. Its behaviour is then governed by one function of time $R(t)$ and one constant k. We call $R(t)$ the scale-factor of the Universe; it can be thought of as giving the *time-dependence* of the distance between two particles (galaxies) and therefore governs the rate of expansion of the Universe. It does not matter which two particles we take because of the uniformity we are assuming. The quantity k in the Newtonian theory gives the total energy, kinetic plus potential, of a particle and so specifies whether the matter in the Universe forms a gravitationally bound system and so whether the expansion continues indefinitely. (In relativity theory k determines the curvature of 3-dimensional space at any one time, and also whether the expansion continues indefinitely.)

There are four types of model universe in which we are mainly interested. The first three are pressure-free:

(i) $k = 0$ (Einstein–de Sitter model). In this case

$$R(t) \propto t^{\frac{2}{3}} \quad \text{(fig. 37)},$$

$$6\pi\, G\rho t^2 = 1,$$

$$t = \tfrac{2}{3}\tau,$$

and
$$\frac{8\pi}{3}\, G\rho\tau^2 = 1,$$

where τ is the Hubble constant (p. 103). The present value of τ is 10^{10} years, and so the present age since the moment of infinite density at $t = 0$ is 6.7×10^9 years, and the present density is 2×10^{-29} g cm^{-3}.

(ii) $k > 0$ (oscillating model). In this case $R(t)$ is a cycloid (fig. 38), the present age is less than in the Einstein–de Sitter model, and the density is greater.

(iii) $k < 0$ (ever-expanding model). In this case $R(t)$ begins like $t^{\frac{2}{3}}$ and ends like t (fig. 39); that is, the model ends up freely expanding with essentially no gravity holding it back. Its age now is greater than in the Einstein–de Sitter model and its density is less. In fact its age comes closer to τ the smaller is the present value of the density (which cannot, of course, be reduced below the density contributed by known galaxies $\rho_g \sim 10^{-30}$ g cm^{-3}).

The fourth model is of interest in the earliest stages of the expansion, when radiation may dominate completely over matter (chapter 12), and radiation pressure must be taken into account.

(iv) Radiation-filled model. In this case

$$R(t) \propto t^{\frac{1}{2}} \quad (t \text{ small}).$$

RED SHIFT

Non-technical books on cosmology usually give a simplified account of the red shift associated with the theoretical model universes. This procedure was reasonable at a time when observed red shifts were very small, but would be misleading now when red shifts as large as 2 are almost commonplace, and when still larger ones may yet be discovered. The same remark applies to other observable properties of a distant source such as its apparent brightness (optical or radio) and its angular size. We are now in the regime where to analyse the observations we need rather accurate estimates of fully relativistic effects.

We shall not give the detailed calculations here but simply quote the results. Two points are worth noting, however. The first is that an object which partakes exactly of the motion of the substratum has constant r, θ, ϕ co-ordinates in the Robertson–Walker metric. These are thus co-moving co-ordinates of a type often adopted in fluid dynamics (where the phrase 'following the motion' is commonly used). The fact that such an object is receding from us is then contained in the function $R(t)$ which must be, at the moment, an increasing function of time. The second point is that distant objects are detected by the electromagnetic radiation (optical, or radio and perhaps X-) that they emit. The path of such radiation through space-time is given in relativity by a line of zero interval between any pair of points along it, that is, by putting $ds = 0$ in the Robertson–Walker metric. This enables us to calculate the time t at which radiation was emitted that we receive now (at time t_0) from a given source.

These considerations lead to the following simple formula for the red shift of a source that emitted its radiation at time t:

$$\frac{\lambda_0}{\lambda} = 1 + z = \frac{R(t_0)}{R(t)}.$$

The observed wavelength λ_0 and the emittted wavelength λ are in the same ratio as the scale-factors of the Universe at the moment of observation and the moment of emission. To relate this result to the Hubble law we consider a nearby source for which t differs little from t_0 ($t = t_0 - \delta t$). Then we can write

$$1 + z = \frac{R(t_0)}{R(t_0 - \delta t)} \sim 1 + \delta t \frac{\dot{R}(t_0)}{R(t_0)}.$$

Since for small z the classical Doppler formula is a good approximation, we have

$$z \sim \frac{v}{c}$$

and so

$$v \sim \frac{c\delta t}{\tau_0}$$

from (4), which is just the Hubble velocity–distance law with distance given by $c\delta t$. By contrast if z is large we cannot use this approximation. In special relativity z is related to v as follows:

$$1 + z = \left(\frac{c+v}{c-v}\right)^{\frac{1}{2}},$$

so a red shift of 2 would correspond to a velocity of 80 per cent of the speed of light. This relation between z and v is sometimes used in describing the significance of a large z, but since the Robertson–Walker metric is not the special relativity one this is really rather misleading (unless, of course, the QSOs are local and their red shifts are straightforward Doppler shifts after all).

If we know only the red shift of an object we do not know very much about it if the function $R(t)$ for the actual Universe is unknown, which is in fact the case. All we can say is that when the radiation was emitted the Universe was more contracted than it is now by a factor $1+z$, so the density in the Universe was then greater than it is now by the factor $(1+z)^3$. For $z = 2$ this factor is 27, which is very substantial. If we knew the time at which the radiation was emitted we could combine this information with the red shift to derive the function $R(t)$. A conceivable way of determining t would be to detect the dispersion introduced into the radio waves from a variable radio source by an ionised inter-galactic gas. This possibility is discussed on p. 145; the outlook at the moment is not promising.

An alternative method is to determine the apparent optical or radio luminosity of the source which would give a measure of its distance and so of its light-time away. However, the red shift itself influences the apparent luminosity (a receding source being fainter than a stationary one of the same absolute luminosity). To deal with this situation it is convenient to introduce the concept of a luminosity-distance.

LUMINOSITY-DISTANCE

If a source of absolute luminosity L has an apparent luminosity l then its luminosity-distance D is defined to be

$$D = \left(\frac{L}{4\pi l}\right)^{\frac{1}{2}},$$

where L and l refer to the total radiation over all frequencies. In practice observations are made with a limited bandwidth and allowance must be made for the spectrum of the source because of the red shift involved. The definition of luminosity-distance is chosen so that the ordinary inverse square law is satisfied in terms of it. It can then be shown that

$$D = R(t_0)\,(1+z)\,\frac{r}{1+\frac{1}{4}kr^2}. \tag{12}$$

Since the co-ordinate r is not directly observable it is convenient to eliminate it using the equation for a light-ray. We then find

$$D = \frac{cT_0}{q_0^2}\{q_0z+(q_0-1)[(1+2q_0z)^{\frac{1}{2}}-1]\}. \tag{13}$$

For small z this reduces to

$$D = zcT_0\{1+\tfrac{1}{2}(1-q_0)z\}.$$

The first term in z gives us back the Hubble law again, as we would expect since for small z luminosity-distance and light-time distance hardly differ. If the observations could be carried through to the higher terms in z we would have a means of determining the present value of the deceleration-parameter q_0 and so of the density ρ and the sign of the curvature k. This observational problem and its difficulties were discussed in chapter 3.

ANGULAR DIAMETERS

Another observational method of distinguishing between the different cosmological models is to study the dependence on red shift of the angular diameters of a class of objects that hopefully have a well-defined linear diameter. At very small distances the angular diameter would, of course, be inversely proportional to the distance, but for sources whose red shift is appreciable important relativistic effects come into play. To determine these effects we put $dr = dt = d\phi = 0$ in the Robertson–Walker metric and solve for $d\theta$:

$$d\theta = a\frac{(1+\frac{1}{4}kr^2)}{rR(t)},$$

where we have written a for ds, the linear diameter of the source, and t is the time when the radiation was emitted. A neater expression for $d\theta$ is obtained if we introduce the luminosity-distance D from (12) and express $R(t)$ in terms of the red shift z. We then obtain

$$d\theta = \frac{a(1+z)^2}{D}.$$

We can regard this as a relation between angular diameter, red shift and q_0 since the luminosity-distance D is known in terms of the red shift and q_0 from (13). In principle, then, we could determine q_0 from the observed $d\theta$, z relation. In practice this is very difficult owing to the intrinsic spread in the linear diameter a for all known classes of source.

Despite this difficulty the $d\theta$, z relation is of great interest. At large z ($z \gg 1$), D is nearly proportional to z and so $d\theta$ is nearly proportional to z. In order words at large z the angular diameter actually *increases* with increasing red shift and luminosity-distance. This is a consequence of the 'bending' of light in the curved space-time of the Robertson–Walker models, which thus have a lens-like behaviour. Of course, at small red shifts the angular diameter decreases as the red shift increases and so the angular diameter goes through a minimum. In the Einstein–de Sitter model, for example, this minimum occurs at a red shift of less than 2, and some QSOs have a greater red shift than this. However, since the QSOs have a large spread in their intrinsic size, it has not yet been possible to exploit this fact.

It must also be stressed that this remarkable behaviour of the angular diameters depends on the assumption that there is a substantial and essentially uniform distribution of matter between the galaxies. If, in fact, most of the matter in the Universe is contained within galaxies, so that the space along a line of sight between us and a QSO is essentially empty, then the angular diameter of the QSO would behave somewhat differently. Perhaps one day this effect will be used to decide whether or not there is a significant intergalactic gas.

NUMBER COUNTS AND OLBERS' PARADOX

If a class of objects such as galaxies or radio sources is distributed uniformly in space at each cosmic epoch, one can calculate from the Robertson–Walker metric the relative number of those objects that have a given red shift or a given measured brightness. We shall not give the relations here because attempts to compare them with observation have been frustrated by selection effects in the case of galaxies and evolutionary effects in the case of radio sources. These evolutionary effects have been discussed in chapter 6; they arise because the fainter, more distant, objects are being seen at earlier epochs, when their intrinsic properties may have been (and probably were) different from their present properties. This effect can easily dominate over the differences between the different Robertson–Walker models. Moreover we do not understand the objects sufficiently well to know how they do in fact evolve. The observations we can make thus contain evolutionary effects and cosmological effects mixed up in unknown proportions.

A further aspect of the number counts that is of great practical importance concerns the contribution of all the sources to the unresolved background level of radiation. If one measures the flux of radiation reaching the earth in a solid angle much larger than that of an individual source, one is measuring this background. Sources that may be too faint to be detected individually can, in the aggregate, make a substantial contribution to the background. Thus to calculate the background it is necessary to know the relative number of sources of different absolute luminosity and to compute their contributions out to rather large red shifts.

This last point is often referred to as Olbers' Paradox. This paradox arose about 200 years ago from an attempt to compute the contribution of all the stars (or as we would now say, the galaxies) to the background light of the night sky. If we forgot about the red shift we would suppose that the number of galaxies at a distance r increases as r^2, while the contribution of each galaxy decreases as $1/r^2$. Thus the contribution to the background from the galaxies at a distance r would be independent of r. Distant galaxies would thus make an important contribution, and indeed there would be convergence difficulties as we let r tend to infinity. These difficulties are resolved by the red shift, which weakens the contribution of distant galaxies over and above the inverse square law. A detailed calculation confirms the intuitively obvious idea that the background is approximately the same as that due to a distribution of sources without red shift that cuts off at a distance of $c\tau_0$, at which the linear Hubble law leads to a velocity of recession equal to the velocity of light c. The background is thus roughly given by

$$nLc\tau_0,$$

where n is the present concentration of sources and L is their absolute luminosity.

The intergalactic starlight density is then found to be about 10^{-2} eV cm^{-3}, which may be compared with the 1 eV cm^{-3} characteristic of our own Galaxy. Thus at a point inside our Galaxy, the other galaxies contribute 1 per cent of the background starlight; at a point well between the galaxies the nearest galaxy no longer makes the dominant contribution. In terms of mass density the intergalactic starlight contributes about 10^{-35} g cm^{-3} to the mean density of the Universe. This is much smaller than the smoothed-out matter density which is at least 10^{-31} g cm^{-3}.

This question of the integrated background is a very important constraint on astrophysical hypotheses about the evolution of sources and also on hypotheses about the emissive power of the intergalactic medium. Such hypotheses must obviously not lead to a background that exceeds the observed background at any wavelength whatever. As we shall see in later chapters, this constraint is of particular relevance at radio and X-ray wavelengths.

COSMOLOGICAL COINCIDENCES

We cannot leave our theoretical discussion of cosmology without mention of the famous numerical coincidences. These coincidences are the starting points of many speculative theories some of which are of great intellectual interest, but in accordance with the pragmatic spirit of this book we shall not describe them here. All we will do is to state the coincidences, and to indicate how they might be explained with the least deviation from accepted theory.

The coincidences are best expressed in terms of dimensionless ratios. One such ratio is the radius of the universe $c\tau_0$ divided by the classical radius of an electron $(e^2/m_e c^2)$ (e and m_e are the charge and mass of an electron). This ratio is a very large number, of order 10^{40}. Another very large number is the ratio of the electric to the gravitational force between an electron and a proton. This ratio $(e^2/G m_e m_p)$ is 2×10^{39}. The near equality of these two numbers constitutes the first coincidence that we must consider.

The second coincidence is that this same large number is approximately equal to the square root of N_0, 'the number of particles in the Universe', that is, the number of hydrogen atoms within a sphere of radius $c\tau_0$, as derived from observations of the mean density ρ_0 of the Universe. We have already stressed that the density ρ_0 and so the number N_0 are uncertain by about a factor of 100. This should not, however, be regarded as detracting from the significance of the coincidences, since *a priori* the two sides of each coincidence could differ by a very large factor indeed.

From these coincidences we can deduce by simple manipulation another remarkable one, this time relating the gravitational constant to the cosmological quantities ρ_0 and τ_0:

$$G\rho_0\tau_0^2 \sim 1.$$

We shall find it convenient to take this as one of the basic coincidences that we shall attempt to explain. We may take for the second one the relation

$$\frac{c\tau_0}{e^2/m_e c^2} \sim N_0^{\frac{1}{2}} \sim 10^{40}.$$

A possible explanation of the first coincidence is based on what is known as Mach's principle. This principle is discussed in detail in my book *The Physical Foundations of General Relativity*. It states in essence that matter has inertia because of its interaction with the rest of the matter in the Universe. According to Einstein this interaction is gravitational in character, and so its strength depends on the value of the gravitational constant G, and on the amount of matter in the Universe. The problem of calculating the total effect of all this matter is rather similar to that of calculating the background radiation due to a uniform distribution of sources, and again we have to introduce a cut-off at a distance of the order of $c\tau_0$. The condition that there is just enough matter in the Universe to induce the observed amount of inertia into a local body then has the form

$$G\rho_0\tau_0^2 \sim 1,$$

which is our first coincidence. A similar argument could have been used when the location of the spiral nebulae was in dispute (p. 37) to indicate that the Milky Way does not in fact comprise all the material in the Universe.

The second coincidence, which we shall use in an astrophysical context on p. 144, can be given a relatively conservative explanation. The most detailed version of this explanation has been developed by Brandon Carter. It has suggestive value only, in that some of its steps are plausible rather than demonstrable, but it does show that there is no definite need at the moment to introduce a theory relating microscopic and macroscopic phenomena in a new way. The basis of the explanation is that life is possible only during a relatively short phase of the evolution of the Universe after the galaxies have formed and when stars are shining in a stable manner. The time at which this phase occurs depends on the microscopic constants and on the gravitational constant, since these determine the details of stellar structure and evolution. This time must also represent the present age of the Universe which, as we have seen, is of the same order as the present value of the Hubble constant τ_0. We thus get a relation between microscopic constants, G and τ_0. Our second coincidence is also of this form if we use the first one to eliminate the ρ_0 hidden in N_0. It therefore seems premature to invoke a far-reaching new theory in order to explain the second coincidence.

IRREGULAR MODELS OF THE UNIVERSE

We must finally consider the relation between the exactly uniform Robertson–Walker models and the actual Universe with all its irregularities that extend out to a scale of at least 1 megaparsec (cluster of galaxies) and perhaps out to 50 megaparsecs (if super-clustering occurs) or even further (p. 96). One important aspect of this problem is how the galaxies themselves managed to form in an expanding universe. Some formation process is required since presumably the Universe as a whole was once very much denser than the material now in galaxies (which have a density $\sim 10^{-24}$ g cm^{-3}). This problem turns out to be very difficult and is still unsolved. All one can say is that if the Universe was originally uniform then the small irregularities that arise from statistical fluctuations would not develop into galaxies in the time available. It appears to be necessary to assume that even in its early stages large density fluctuations were present. Whether such large fluctuations could arise spontaneously or whether they are a consequence of a previous collapse phase of the Universe that preceded the present expansion phase is unknown.

These questions are related to another difficult problem, that of the singular moment at $t = 0$ in the Robertson–Walker models when the density of the Universe was infinite. It has often been suggested that this singularity need not trouble us because it may be a consequence of the artificially exact symmetry assumed in these models. If the galaxies move exactly radially according to the Hubble law, it is argued, it is not surprising that all the material in them was once at the same place at the same time. By contrast in an irregular universe no singularity may have existed. This argument is not correct. It has been shown by S. W. Hawking, G. F. R. Ellis and R. Penrose that so long as the material of the Universe obeys a physically reasonable equation of state, then according to general relativity there must have been one or more physical singularities in our past. The essential reason is that in general relativity self-gravitation is so strong that even in our irregular universe some at least of the material must have been squeezed to infinite density. Quantum mechanical considerations might possibly enable us to avoid a literal singularity but apparently they cannot prevent the density getting very high indeed, say 10^{59} g cm^{-3} which corresponds to a radius of curvature of 10^{-26} cm. For practical purposes such a density might well be regarded as a singularity.

More work will have to be done to discover the detailed nature of the singularities in an irregular Universe. There are indications that they may be well separated from one another and that most of the material in the Universe does not in fact pass into or out of them. If this is so a 'bounce' from a previous contracting phase might become a theoretical possibility. The question whether the actual Universe experienced such a bounce is in many ways the most exciting problem in cosmology. It is also one of the most difficult.

9 The search for intergalactic atomic hydrogen

INTRODUCTION

We saw in chapter 8 that amongst all the relativistic models of the Universe a special role is played by the Einstein–de Sitter model in which

$$\frac{8\pi}{3} G\rho\tau^2 = 1.$$

The observed values of G and τ then imply for this model that at present

$$\rho \sim 2 \times 10^{-29} \text{ g cm}^{-3}.$$

If we assume that most of the material is in the form of hydrogen we would then have for the present particle density

$$n_{\text{H}} \sim 10^{-5} \text{ cm}^{-3}.$$

By contrast the amount of matter that has been observed so far, namely in the form of galaxies, gives a particle density of only

$$n_{\text{H}} \sim 10^{-6} \text{ cm}^{-3}.$$

If this latter density is close to the actual total density, then the Universe would be open and the kinetic energy of its expansion would greatly exceed its gravitational energy. It would clearly be of considerable interest to know which particle density is nearer the truth, and great efforts have been made to decide this question. There may exist a large number of faint galaxies whose contribution to n_{H} has not been included in the above estimate but this possibility has usually been discounted. Attention has been focused instead on the possible contents of intergalactic space.

In considering this question we must immediately make a distinction between intergalactic space within a cluster of galaxies and intergalactic space between clusters of galaxies. We may recall that many of these clusters would be rapidly dispersing unless they were stabilised by intergalactic matter that contributed from 10 to 30 times more material than the galaxies themselves. Such matter has not yet been detected (except between occasional pairs of obviously interacting galaxies) but it could be in a form that would hitherto have escaped detection (for example, as ionised hydrogen at a temperature of $10^6 \,^\circ$K). Even if it exists it is not clear whether it could contribute as much as 10^{-5} particles per cm^3 to the overall density. The estimates would not be precise enough to decide this.

It is therefore an important question to determine whether intergalactic space as a whole contains a significant amount of material. This is a difficult question *a priori* because a general density of 2×10^{-29} g cm^{-3} would not yet have been observed if it were in the form of individual faint stars, rocks, neutrinos or gravitational waves. In this book we shall restrict ourselves to the possibility that the 'missing material', if it exists, is in gaseous form. This point of view has the advantage of keeping the discussion in contact with the astrophysical problem of the formation and evolution of galaxies and galaxy clusters. It corresponds to the fact that we do not know how efficient the process of galaxy formation is.

The likely chemical composition of the gas is the first question we must consider. Judging by the composition of galaxies, most of it is probably in the form of hydrogen. What heavy elements are present depends on when and where these elements are formed, a problem that is discussed in chapter 11. We may summarise the results of that discussion by saying that it is quite likely that 27 per cent by mass of the material is helium and that a negligible fraction is in the form of heavier elements. No way has yet been found of detecting the helium but if present it could have a significant effect on the cooling rate of the intergalactic gas. On the other hand, intergalactic hydrogen of particle density $\sim 10^{-5}$ cm^{-3} does produce effects that are potentially detectable, and this is the question we mainly discuss in this chapter and the next. The hydrogen has rather different effects if it is atomic or ionised, and it is convenient to take these two cases separately. Accordingly we consider atomic hydrogen in this chapter and ionised hydrogen in the next.

21-CENTIMETRE EFFECTS

An intergalactic distribution of atomic hydrogen would act both as an absorber and an emitter of 21-cm radiation. The absorption was first looked for by G. B. Field in 1958. Field studied the spectrum of the radio source Cygnus A, the brightest extragalactic radio source in the sky, in the neighbourhood of 21 cm. Since intergalactic hydrogen would absorb only in a very narrowly defined wavelength range we must take into account the red shift associated with the expansion of the Universe. Clearly the hydrogen near us would absorb at 21 cm, but the hydrogen near Cygnus A would, from our point of view, absorb at a wavelength of $21(1+z)$ cm, where z is the red shift of Cygnus A. Intermediate hydrogen would appear to absorb at an intermediate wavelength, and so we would expect to see an *absorption trough* stretching between 21 cm and $21(1+z)$ cm. Of course the failure to observe such a trough to a certain precision would enable us to place an appropriate upper limit on the density of intergalactic atomic hydrogen.

This concept of an absorption trough will recur later in this chapter with regard to other absorption processes. There is, however, one respect in which the 21-cm action of atomic hydrogen is rather subtler than that of the other examples of absorption we shall meet. It arises from the fact that the energy gap between the two (hyperfine) states involved in the 21-cm transition is very small (corresponding to the fact that 21 cm is a very long wavelength for an atom to emit or absorb). The energy gap is only 5×10^{-6} electron volts. It is useful to express this energy E as a temperature T by the relation $E = kT$, where k is Boltzmann's constant. The resulting temperature is only 0.06 °K. By contrast the effective temperature of the intergalactic radiation field at 21 cm is likely to be a good deal greater than this. According to the recent measurements of T. L. Howell and J. R. Shakeshaft it is about 3 °K (chapter 14). Thus from the point of view of the two energy levels involved the hydrogen is immersed in a heat bath at a very high temperature. It will then no longer be true that the vast majority of the hydrogen atoms are in the lower of the two states (the ground state). In fact there will be nearly as many atoms in the upper state as in the lower.* In these circumstances a process called stimulated emission plays an important part, and complicates the interpretation of the observations.

Because of this complication the depth of the absorption trough does not give directly the density n_H of atomic hydrogen, as it would if most of the hydrogen were in its ground state and stimulated emission were unimportant. It gives us instead n_H/T_s, where T_s, the so-called spin temperature, measures the relative number of atoms in the two states. In fact Field failed to observe an absorption trough to a precision which implied that

$$\frac{n_H}{T_s} < 4 \times 10^{-7} \text{ cm}^{-3} \text{ deg}^{-1}.$$

Later observations by Field reduced this limit by a factor 3, and the most accurate observations to date, by A. A. Penzias and E. H. Scott give

$$\frac{n_H}{T_s} < 2.3 \times 10^{-8} \text{ cm}^{-3} \text{ deg}^{-1}.$$

To interpret this result we need to know the value of the spin temperature T_s. If it were determined by the background radiation it would be 3 °K and we would have $n_H < 7 \times 10^{-8} \text{ cm}^{-3}$, which would clearly be an important result. However, a number of complicated processes contribute to the value of T_s. The most important of these is the absorption of Lyman α radiation (see p. 133) by a hydrogen atom in one of the 21-cm levels followed by emission of Lyman α leaving the atom in the other of the 21 cm levels. The Lyman α from galaxies and from QSOs would probably lead to a value of about 4 °K for T_s, giving $n_H < 9 \times 10^{-8} \text{ cm}^{-3}$, much the same result as before. However if the intergalactic gas is partially ionised it would produce its own Lyman α. If $n \sim 10^{-5}$ cm^{-3}, T_s might be as large as 80 °K, in which case we would have $n_H < 2 \times 10^{-6} \text{ cm}^{-3}$. Thus in all circumstances the density of atomic hydrogen is probably somewhat less than the critical value of 10^{-5} cm^{-3}. Owing to the observational uncertainty in the value of Hubble's constant (and therefore of the critical density) this is not quite adequate to rule out an Einstein–de Sitter model made of atomic hydrogen.

* For simplicity we here neglect the degeneracy of the upper state.

Fortunately the emission at 21 cm is easier to interpret. The emission intensity depends on the number of atoms in the upper state and normally this would depend sensitively on the spin temperature. However in our case the spin temperature is so high that, as we have seen, we may assume that there are as many excited as unexcited atoms. Hence the observed emission intensity hardly depends on the spin temperature, and gives us n_H directly (so long as the spin temperature exceeds the 3 °K background temperature at 21 cm). Because of the red shift the emission is spread out longwards of 21 cm, and since it is superposed on many other sources of continuous emission, the observers look for a step at 21 cm where the emission from the intergalactic gas would begin. No step has been observed to the precision of the measurements which leads to the limit (Penzias and Wilson)

$$n_H < 3 \times 10^{-6} \text{ cm}^{-3}$$

which, in view of the uncertainty in the value of the Hubble constant, is barely adequate to rule out an Einstein–de Sitter model made of atomic hydrogen.

X-RAY ABSORPTION

It has been known since 1962 that a roughly isotropic flux of X-rays is incident on the Earth. This background was discovered by Giacconi, Gursky, Paolini and Rossi in their first rocket experiment looking for celestial X-rays, the experiment in which they also discovered the famous discrete X-ray source Scorpio X-1. The intensity of the background has now been measured at several wavelengths, and its spectrum is shown in fig. 46. The origin of this background is not known although it seems to be generally agreed that most of it comes from outside our Galaxy. Some pro-posed mechanisms will be mentioned later (p. 189). If the longest wavelength (lowest energy) X-rays come from the Universe as a whole, as some of these mechanisms would imply, the possibility of the absorption of the X-rays by intergalactic gas becomes an important consideration.

This absorption would occur when an X-ray photon ionises a hydrogen or helium atom. The emitted electron may eventually recombine with a proton or α-particle to form another hydrogen or helium atom, but it is likely to do so in many steps, each of which releases a less energetic photon than the original X-ray. This would be a case of true absorption in that it results in a reduction in the number of photons with the original energy. By contrast the 21-cm 'absorption' is really a scattering effect since each 21-cm photon that is absorbed by a hydrogen atom is then re-emitted as a 21-cm photon. It is, however, usually re-emitted in a different direction, so that it is lost to the original line of sight. Thus the 21-cm scattering would be unimportant if one were looking at a diffuse background, but for a discrete source like Cygnus A scattering out of the beam would have the same effect as true absorption; that is, there would be a reduction in the intensity of the source.

Now a photon that can just ionise a hydrogen atom in its ground state would have a wavelength λ_0 of 912 Å. The opacity of atomic hydrogen for X-rays of wavelength λ less than λ_0 is proportional to $(\lambda/\lambda_0)^3$. Our best chance of observing the absorption clearly lies with looking at the longest wavelength X-rays possible.

The spectrum in fig. 46 shows no clear signs of absorption down to the longest wavelengths observed, about 50 Å. If this radiation comes from the Universe as a whole we may set the following upper limit on the concentration of intergalactic atomic hydrogen and helium:

$$n_{\mathrm{H}} < 10^{-7}\ \mathrm{cm}^{-3},$$

$$n_{\mathrm{He}} < 3 \times 10^{-9}\ \mathrm{cm}^{-3}.$$

If we assume that $n_{\mathrm{He}}/n_{\mathrm{H}}$ is about 0.1 (chapter 13) we can conclude that
$$n_{\mathrm{H}} < 3 \times 10^{-8}\ \mathrm{cm}^{-3},$$

a stronger limit than we obtained from 21-cm effects, but one depending on the uncertain assumption that the 50 Å X-rays come from the Universe as a whole (see p. 142).

LYMAN α ABSORPTION

Our discussion up to this point has shown that the most powerful way of detecting absorption by intergalactic neutral hydrogen would be to observe at a wavelength λ that is precisely characteristic of an absorption process in the hydrogen atom. Such a resonance condition would ensure that the absorption was at a maximum. It would also be advantageous to have most of the atoms in the lower of the two states involved so that there would be no stimulated emission to reduce the absorption. Since most inter-galactic hydrogen atoms would be in their ground state (apart from the 21-cm splitting), the transitions concerned would have to be those from the ground state, that is the Lyman series, starting with Lyman α at 1216 Å and culminating in the ionisation of the atom at the Lyman edge at 912 Å. The difficulty is that in the wavelength range 912–1216 Å the opacity not only of the Galaxy but also of the Earth's atmosphere is very large.

Recently this problem has been ingeniously solved by Scheuer and by J. E. Gunn and B. A. Peterson. They make use of the large red shift ($\sim$ 2) of some of the QSOs (assumed here to be cosmo-logical in origin). We would expect a Lyman α absorption trough stretching between 1216 Å and 1216(1 + z) Å in the spectrum of a QSO with a red shift z, the absorption at 1216 Å occurring near the Galaxy and that at 1216(1 + z) Å occurring near the QSO. For $z \sim$ 2 the trough would stretch right into the visible at about 3600 Å. In fact the Lyman α line itself is observed in emission in all QSOs with $z \sim$ 2, and the absorption trough should stretch from this line to the lowest wavelength usually observed, namely $\sim$ 3000 Å. In practice the observers look for a depression in the continuum across the emitted Lyman α line. Gunn and Peterson originally believed that a 40 per cent depression did occur in the spectrum of 3C 9, but later work on that spectrum and on the spectra of other QSOs with $z \sim$ 2 have failed to detect a significant depression (that is, it is less than $\sim$ 20 per cent) (fig. 43). This leads to the following limit on n_{H} near the QSO:

$$n_{\mathrm{H}} < 3 \times 10^{-11} \text{ cm}^{-3} \quad \text{at} \quad z \sim 2,$$

and if we assume that $n_{\mathrm{H}} \propto (1 + z)^3$ (see chapter 8), to the following limit for the present value of n_{H}:

$$n_{\mathrm{H}} < 10^{-12} \text{ cm}^{-3},$$

a remarkable testimony to the value of working at resonance and in the absence of stimulated emission.

This upper limit is seven powers of ten less than the value of the particle density we have been testing for, and it might seem to settle quite decisively the question of whether an intergalactic gas exists. However, in a sense, the Lyman α absorption test proves too much. For it is perhaps unlikely that the formation of galaxies and intergalactic gas clouds would be so efficient a process that substantially less than, say, 1 per cent of the material in the Universe is

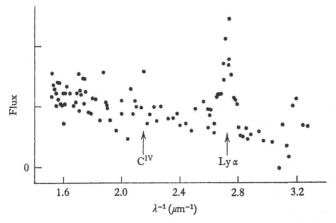

Fig. 43. The photoelectric spectrum of the QSO 3C9, which has a red shift of 2.012. Note the absence of a depression in the continuum on the short wavelength side of the red-shifted Lyman α line. (After E. J. Wampler, *Astrophys. J.* **147**, 1 (1967).)

left over as a widely distributed intergalactic gas. Even on the conservative basis of taking for this fraction 10^{-3}, the intergalactic density would be $\sim 10^{-9}$ cm^{-3}, which is still a thousand times greater than the upper limit. This suggests that we should look elsewhere for an explanation of the lack of Lyman α absorption. One cannot suppose that the hydrogen is in molecular form, because again one would expect absorption effects that have not been observed. The resulting limit on the concentration of hydrogen molecules is 10^{-8} cm^{-3}. The natural assumption to make then is that the gas is sufficiently highly ionised for the Lyman α absorption to be negligible. In point of fact this high degree of ionisation had been suggested, on other grounds, before the Lyman α observations had been made. The problem of detecting an intergalactic gas would thus become the problem of detecting ionised gas, and, of course, of understanding how the gas comes to be ionised. These questions will be discussed in the next chapter. Before leaving the atomic case, however, we would like to mention some interesting variants of the Lyman α method which may lead to the detection of heavy elements in intergalactic space and of atomic hydrogen and heavy elements in clusters of galaxies.

ABSORPTION BY HEAVY ELEMENTS IN INTERGALACTIC SPACE

According to current theories (chapter 13) the elements heavier than hydrogen and helium are formed in galaxies, and so would be expected to have a smaller abundance relative to hydrogen in intergalactic space than in galaxies. Nevertheless the absorption methods are so sensitive that it might be possible to detect even a small trace of heavy elements. Moreover the processes that ionise the hydrogen, leaving it incapable of resonance absorption, may produce ions of heavy elements that still contain several electrons, and so could still absorb somewhat like a hydrogen atom. Since the probability of resonance absorption is not much less for such ions than it is for hydrogen, these ions could be detected if their concentration exceeded $\sim 10^{-12}$ cm^{-3}, so long as the wavelength of the transition is brought into the visible by the red shift of the source.

This latter condition would not be satisfied by singly ionised helium, and absorption by this ion must be sought from above the atmosphere where there is little ultra-violet absorption. The most likely candidates on the basis of their high abundance in the Galaxy and of their atomic properties are triply ionised carbon, that is C^{IV} ($\lambda_0 = 1548$ Å), N^V ($\lambda_0 = 1238$ Å) and O^{VI} ($\lambda_0 = 1031$ Å), and of these the most favourable is O^{VI} (although it needs a red shift rather greater than 2 to bring the edge of the absorption trough into the visible). The possibility of detecting O^{VI} is indicated in fig. 44, for a range of possible temperatures of the intergalactic gas and for two possible hydrogen densities. There appears to be a marginal chance of success for the QSOs with the largest red shifts.

ABSORPTION BY INTERGALACTIC GAS IN CLUSTERS OF GALAXIES

We have already seen (p. 129) that clusters of galaxies may contain intergalactic gas of density in the range 10^{-3}–10^{-5} cm^{-3}, although this gas has not yet been detected. Now the line of sight to a QSO with a red shift ~ 2 probably passes through several clusters, most

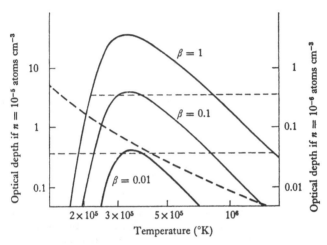

Fig. 44. The optical depth expected on the short-wavelength side of the O^{VI} absorption edge in the spectrum of a distant source plotted as a function of the intergalactic gas temperature (on the simplifying assumption that the temperature alone determines the ionisation balance). The oxygen has a fraction β of its local abundance relative to hydrogen. The horizontal lines indicate the minimum optical depth that would result in a detectable absorption edge. The dashed curved line shows the optical depth due to Lyman α absorption by neutral hydrogen. (After M. J. Rees and D. W. Sciama, *Astrophys. J.* **147**, 353 (1967).)

of them too faint to be detected optically. If there is any significant amount of atomic hydrogen, C^{IV}, N^{V} or O^{VI} in these clusters, an absorption *line* would be produced which would be red shifted by the red shift of the cluster and might thereby be brought into the visible. We would then be faced by the intriguing situation of having an observable absorption line or lines produced by a cluster of galaxies too faint to be seen directly.

The possibilities for H^I, C^{IV} and N^V, all of which could be observed with the present red shifts, are indicated in fig. 45. Absorption lines have been detected in many QSOs, especially those with large red shift, but as we have seen, there is at the moment great confusion about their interpretation. The absorption lines identified so far probably arise in the outer envelopes of the QSOs

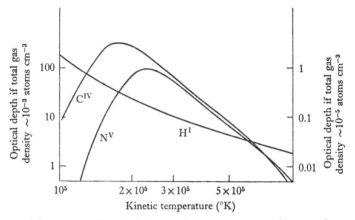

Fig. 45. The expected optical depth of the Lyman α, N^V (1238 Å) and C^{IV} (1548 Å) absorption lines due to a gas cloud of radius 1 Mpc with turbulent or expansion velocities of 500 km s^{-1}. It is assumed that the C and N have their local abundance relative to H. (From D. W. Sciama and M. J. Rees, *Nature* **212**, 1001 (1966).)

themselves. However the study of absorption lines is in its infancy and this technique may well become a useful method of studying the properties of the intergalactic gas in clusters of galaxies. Even a well-defined negative result would give one valuable information, a common phenomenon in cosmology.

10 The search for intergalactic ionised hydrogen

INTRODUCTION

We saw in the last chapter that the density of intergalactic atomic hydrogen is far less than any likely value for the density of hydrogen itself. This forced us to the view that the hydrogen is probably highly ionised. We must now ask two questions: (i) why is it ionised, and (ii) what is its density, and, in particular, can it have the critical value of 10^{-5} cm^{-3}? Further questions arise when we consider the general physical state of the intergalactic medium. We have learnt from the study of our own Galaxy that ionised gas on a cosmical scale tends to be permeated by a magnetic field. Is there an intergalactic magnetic field, and if so, what are its effects? The comparison with our Galaxy also leads us to ask whether there is a significant flux of cosmic rays in intergalactic space, and if so again, what are its effects?

The answers to most of these questions are highly tentative today. Nevertheless they are worth discussing in this book because some answers that look in themselves highly reasonable would lead to observable effects that have not in fact been observed. One can therefore limit the possibilities in an interesting way. Moreover the discussion shows how further observations could lead to a positive discovery or to more stringent limits still. In particular it seems quite possible that the existence of an intergalactic particle density of 10^{-5} cm^{-3} may be decided in the near future. The great cosmological significance of such a result has already been underlined.

THE THERMAL HISTORY OF INTERGALACTIC HYDROGEN

If an intergalactic gas exists and is ionised it is probably hot ($T > 10^5$ °K), both because a cold ionised gas would recombine very rapidly and because most ionising mechanisms tend to heat the gas as well. This is not true for all such mechanisms, but for simplicity we shall assume it here. The question why the gas is ionised then becomes the question why it is hot.

At this point we apparently make contact with the idea, discussed more fully in chapter 12, that the early dense stages of the Universe were probably very hot. However what must be stressed immediately is that the origin of that heat is *not* the origin of the heat in the gas at present. The reason is that in the early stages of the Universe the gas must have been in thermal equilibrium with radiation, the hot ionised gas and the radiation being closely coupled together through their mutual electromagnetic interactions. As the Universe expanded the temperature dropped, and when it reached a few thousand degrees (at a time in the past corresponding to a red shift relative to us now of about a thousand), the hydrogen and helium recombined. The resulting paucity of free electrons broke the strong coupling between matter and radiation, each of which then cooled by expansion in an independent manner. The key point here is that the matter must have cooled more rapidly than the radiation, because it has a higher ratio of its specific heats (see chapter 12). But the present value of the radiation temperature does not exceed 3 °K (chapter 14) and so according to this argument the present temperature of the gas must be far less than 3 °K, in complete contradiction to our previous conclusion.

It follows from this that the gas must have been re-heated by some process. What that process was we can only speculate, but it seems natural to relate it to the fact that at some stage during the expansion of the Universe the galaxies and radio sources were formed. These objects eventually produced ultra-violet radiation and cosmic rays, some fraction of which must have escaped into intergalactic space. Rough calculations suggest that this fraction would have sufficed to heat the gas to a temperature of, say, a million degrees. One is reminded of the problem of the heating of the Sun's corona to its observed temperature of a million degrees. The basic point is that in cosmical conditions it is not difficult to heat a dilute gas to a high temperature, in the sense that the energy input required is not large by cosmical standards. The problem therefore is not so much to find an adequate source of energy but rather to discover which of the many possible sources is the dominant one. Thus the high temperature we require of the intergalactic gas is not a problem of principle but rather one of practice.

It would be very interesting to know at what point during the expansion of the Universe the heating began and how long it lasted, or even whether it is continuing to-day. Unfortunately we do not even know at what point the galaxies started to form. What we can do, however, is to place limits on how early the heating could have begun. The main point is that if the gas was re-heated very early then it would have become hot when it was relatively dense. This would have had two consequences. The first is that the radiation field that survived from the earliest stages would have interacted strongly with the hot gas and so would have had its spectrum distorted. The undistorted spectrum would now be that of a black body at 3 °K (chapters 12 and 14). The distortion would occur in the sense of mimicking a higher temperature corresponding to that of the hot gas. Accordingly the distorted spectrum would peak at a shorter wavelength than the 1 mm or so that corresponds to 3 °K. The necessary observations to test whether this distortion actually occurs at submillimetre wavelengths will have to be carried out from above the earth's atmosphere. Such observations should soon be made.

The second consequence of an early re-heating of the gas is that it would itself radiate at a high rate, since the radiation arises from the interaction of the electrons with the protons, and so increases as the square of the density. This radiation would reach us now suitably red shifted, but whereas its effective temperature is proportional to $(1+z)$ (chapter 12), the particle density is proportional to $(1+z)^3$ and so the radiation rate to $(1+z)^6$. Thus the higher the z at which a given gas temperature is reached the higher the intensity we now observe. This intensity must not exceed the observed background at any wavelength. With a present particle density of 10^{-5} cm^{-3} or even 10^{-6} cm^{-3} this becomes a real restriction, the dangerous wavelengths being in the radio and X-ray regions. In order to avoid too large a background at about 20 cm (the most critical radio wavelength) the gas can only be reheated to substantial temperatures (in the range 10^5–10^6 °K) at red shifts less than about 100.

The X-ray background presents a more interesting problem. The early observations of this background were limited to wavelengths less than a few ångströms, that is, to energies exceeding a kilovolt or so. The observed intensity of the background at these energies is incompatible with an intergalactic gas of density 10^{-5} cm^{-3} unless its temperature is less than about 10^7 °K. At lower temperatures only very few electrons would have a kilovolt of kinetic energy, and so only very few could produce kilovolt X-rays. On the other hand, unless the temperature exceeds about 3×10^5 °K the ionisation level would be too low to account for the lack of Lyman α absorption (for $n_\mathrm{H} \sim 10^{-5}$ cm^{-3}). Thus it was predicted that for a gas of this density the temperature would have to be in the general vicinity of 10^6 °K. Now such an intergalactic gas would produce a detectable flux of X-rays at say 0.25 keV, or 50 Å. In the last year measurements have been made at this low X-ray energy (fig. 46). and the observed flux is indeed comparable with the predicted flux. Unfortunately, this does not settle the matter because the soft X-ray flux may have a different origin. All we can say is that the observations are compatible with the existence of a hot intergalactic gas of density 10^{-5} cm^{-3}. The gas has, perhaps unexpectedly, overcome a severe hurdle. But it is not home yet.

OTHER EFFECTS OF AN IONISED INTERGALACTIC GAS

We now consider other effects which an ionised intergalactic gas of density 10^{-5} cm^{-3} would have in the hope that one day they will be subjected to observational test.

(i) *Absorption of low radio frequencies.* The absorption coefficient of an ionised intergalactic gas for radio waves of frequency ν is proportional to $n^2/(\nu^2 T^{\frac{3}{2}})$, where T is the temperature of the gas. It is possible that the background at a few MHz is of extragalactic origin, the integrated effect perhaps of a large number of radio sources. In that case the spectrum of the background would enable us to place limits on the quantity $n^2/T^{\frac{3}{2}}$.

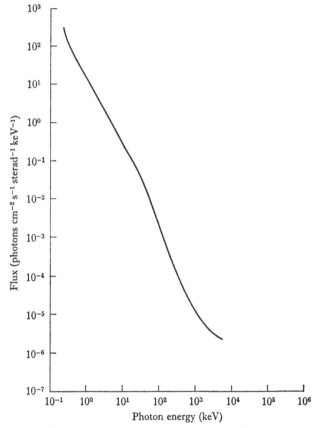

Fig. 46. The energy spectrum of the isotropic X-ray background.

In a similar way we can study the spectra of individual extra-galactic radio sources to see whether they show absorption at low frequencies. Some of these spectra show no signs of absorption down to 10 MHz. (Observations at lower frequencies have usually to be made by rocket or satellite above the ionosphere.) We thus obtain an upper limit on the value of $n^2/T^{\frac{3}{2}}$. For $n \sim 10^{-5}$ cm^{-3} we find $T > 250$ °K, which is not a very strong restriction, but would rule out a very cold ionised plasma. These observations will presumably be extended down to about 1 MHz by future satellite observations. At such a low frequency, absorption in our own Galaxy would be important but could perhaps be allowed for from other observations. In that case the more interesting lower limit of 5000 °K would follow from the absence of intergalactic absorption (for $n \sim 10^{-5}$ cm^{-3}).

(ii) *Thomson scattering.* To see the importance of Thomson scattering we recall the well-known numerical coincidence (chapter 8) that the radius of the Universe $c\tau$ divided by the classical radius of an electron e^2/mc^2 is roughly equal to the square root of the 'number' N of particles in a universe with $n \sim 10^{-5}$ cm^{-3} (that is, in a sphere of radius $c\tau$):

$$\frac{c\tau}{e^2/mc^2} \sim N^{\frac{1}{2}} \quad (\sim 10^{40}).$$

Writing $N \sim \frac{4\pi}{3} n(c\tau)^3$, we have

$$\frac{(c\tau)^2}{(e^2/mc^2)^2} \sim N \sim \frac{4\pi}{3} n(c\tau)^3,$$

or

$$c\tau \sim \frac{1}{n \dfrac{4\pi}{3} (e^2/mc^2)^2}.$$

Since the cross-section σ for Thomson scattering is $\frac{8}{3}\pi(e^2/mc^2)^2$, this coincidence implies that the mean free path $1/n\sigma$ of a photon in the intergalactic gas is comparable with the Hubble radius. Sources with large red shift z would thus be seriously affected.

An exact calculation shows that for $n \sim 10^{-5}$ cm^{-3} (Einstein–de Sitter model) the brightness of a source of red shift z would be reduced by the factor e$^{-\tau_c}$, where the optical depth τ_c is given by

$$\tau_c = 0.046\{(1+z)^{\frac{3}{2}} - 1\}.$$

For a QSO with $z \sim 2$, τ_c has the value ~ 0.2 and e$^{-\tau_c} \sim 80$ per cent, so the effect is beginning to be important at this red shift.

For $z \sim 4$, $\tau_c \sim 0.47$, which would be large enough to affect the source counts. Note that the effect is independent of frequency and so is equally important for optical and radio sources. It is also independent of temperature, and so would give n directly if it could be observed.

It is important to remember that the Thomson effect is a scattering and not a true absorption. Thus the contribution of discrete sources to the total background is unaffected by Thomson scattering. It might be possible to use this fact to limit the particle density n when enough large red shifts are known for radio sources. The reason is that a source of measured flux density S makes a contribution Se^{τ_c} to the background, and too large a value of τ_c might lead to a total estimated contribution exceeding the observed background.

(iii) *Dispersion.* When radio waves propagate through an ionised gas, the speed of propagation depends on the frequency of the wave, that is, the medium is dispersive. Waves of frequency ν ($\nu \gg \nu_p$) propagate with a time lag $\sim \frac{1}{2}\nu_p^2/\nu^2$ per unit time, when compared with propagation in free space. Here ν_p is the plasma frequency $(ne^2/\pi m)^{\frac{1}{2}} \sim 30$ Hz for $n \sim 10^{-5}$ cm^{-3}. For $\nu \sim 20$ MHz the fractional delay $\sim 10^{-12}$. The total propagation time for a source of large red shift $\sim 10^{10}$ years, so the delay at 20 MHz would be about 3 days. More local ionised regions would be expected to produce much less dispersion than this. Thus if a radio source varied on a time-scale not much longer than 3 days, it might be possible to detect the intergalactic gas.

This idea was first worked out by F. T. Haddock and the author in July 1964, when it was thought unlikely that sources would vary in the radio region on the required time-scale, although the corresponding optical variations had been found for some QSOs. However, extragalactic radio variables are now known (p. 74) but they vary only at high frequencies ($\nu \sim 1000$ MHz) where the dispersion is very small, and on too long a time-scale ($\sim$ years rather than days). In case radio variables with the required properties are ever found (for example, very distant pulsars), we record here the formula*

---

* The formula given in the original paper is incorrect. This error was discovered and corrected by Steven Weinberg.

for the time-lag t for a source of red shift z, with $n \sim 10^{-5}$ cm^{-3} (Einstein–de Sitter model)

$$t = \frac{2.7 \times 10^{20}}{\nu^2} \{(1+z)^{\frac{1}{2}} - 1\}.$$

We may remark wistfully that if the effect were discovered and QSOs were found to satisfy this relation it would establish the cosmological nature of their red shift (although allowance would have to be made for dispersion and other frequency dependent effects occurring within the sources). More generally, if n were not 10^{-5} cm^{-3} and the Einstein–de Sitter model were incorrect, the observed t, z relation would give us directly the actual dependence of the scale factor R on time.

(iv) *Faraday rotation.* Many extragalactic radio sources are linearly polarised and the ionised intergalactic gas would contribute to the observed Faraday rotation of their plane of polarisation (p. 34). The size of this effect depends on the magnitude and scale of the intergalactic magnetic field, which are discussed in the next section. However, it is worth noting here that with $n \sim 10^{-5}$ cm^{-3}, $H \sim 10^{-7}$ gauss (the value advocated especially by V. L. Ginzburg and S. I. Syrovatsky) and a scale length $\sim 10^6$ light-years, the gas would produce a Faraday rotation of ~ 40 radians at a wavelength of 1 metre for a source with $z \sim 1$. Typical observed rotations in directions where the effect of the Milky Way is minimised are much less than this, and the absence of any marked dependence of rotation on red shift suggests that if $n \sim 10^{-5}$ cm^{-3} the intergalactic magnetic field is unlikely to exceed about 10^{-8} gauss (or that it is much more tangled than a scale-length of 10^6 light-years would suggest).

This completes our discussion of possible effects arising from ionised intergalactic gas. Some of them are rather tantalising, as they are on the verge of being important. Perhaps by the time this book appears some of them will be promoted from the potential to the actual, a process which has become surprisingly rapid in cosmology in the last few years.

COSMIC RAYS AND MAGNETIC FIELDS IN INTERGALACTIC
SPACE

It may seem premature to discuss cosmic rays and magnetic fields
in intergalactic space inasmuch as there is no direct observational
evidence yet of their existence. Nevertheless we know that cosmic
rays must be leaking out of galaxies on a time-scale small compared
with the Hubble age of the Universe (p. 26), so that the question of
their flux in intergalactic space automatically arises. Moreover the
observed absence of certain effects enables us to place useful limits
on some of the parameters that specify the physical state of the
intergalactic medium. Accordingly we shall discuss these questions
here.

It must be admitted right away that we do not know where the
cosmic rays in our own Galaxy originated. We are not even certain
that their sources are in our Galaxy, although this is the most
likely possibility for cosmic ray energies below about 10^{17} eV
(p. 36). Nevertheless some astronomers have urged that all the
cosmic rays may be universal, that is, that their flux in intergalactic
space may be the same as that observed at the Earth. There is no
direct evidence against this hypothesis. However, if the density of
the intergalactic gas does turn out to be 10^{-5} cm^{-3} the cosmic rays
could not be universal. The reason is that they would interact with
the gas, heating it to a temperature so high that the resulting X-ray
flux would exceed the observed X-ray background. In addition the
interaction would produce π^0 mesons which would decay into
γ-rays, and the resulting γ-ray flux would exceed the upper limit
that has been placed on the γ-ray background in the energy range
50–100 MeV. Since the intergalactic gas density may turn out to
be significantly less than 10^{-5} cm^{-3} this argument could be irrelevant, but it illustrates the way negative results can be used to place
interesting limits on the degree of activity occurring in intergalactic
space.

About the electron component of the cosmic rays we can be
more definite. If their flux in intergalactic space were the same as
that observed at the Earth then their (Compton) interaction with
intergalactic starlight and microwave photons would lead to an

unacceptably high X-ray flux, as explained on p. 189. This argument limits the intergalactic flux of cosmic ray electrons to 10^{-3} of their local flux. Moreover if they achieve this limiting flux the intergalactic magnetic field strength cannot exceed 2×10^{-7} gauss. Otherwise their synchrotron radiation would exceed the observed background radio radiation.

This brings us to the general question of whether there is an intergalactic magnetic field. Some cosmologists have appealed to such a field in order to account for the magnetic fields found in galaxies and radio galaxies, which otherwise would be hard to account for. The suggestion is that the intergalactic field was formed early in the evolution of the Universe in physical conditions quite different from those obtaining in galaxies to day. Then when these galaxies first formed by condensation from the intergalactic gas the magnetic field would have been already present. One can even speculate about the possible differences visible to-day between galaxies whose initial magnetic fields were more nearly parallel or more nearly perpendicular to their rotation axes. It has, for instance, been suggested that the development of a radio galaxy depends on the initial angle between its rotation axis and its magnetic field.

If there is an intergalactic magnetic field it would be important in other contexts too. We have already mentioned synchrotron radiation by intergalactic relativistic electrons, and Faraday rotation of the plane of polarisation of distant radio sources. In addition such a field could play an important part in the propagation of cosmic rays of energies 10^{17} eV which may well be universal. All these matters are speculative, but it seems likely that in the next few years it will be possible to reach definite conclusions about the astrophysical role of intergalactic cosmic rays and magnetic fields. This in turn would have important cosmological implications, especially as an intergalactic magnetic field may be of critical importance in the early dense stages of the expanding Universe.

11 The helium problem

INTRODUCTION

The helium problem is part of a larger problem, namely to understand the abundances of all the chemical elements relative to hydrogen in all parts of the Universe. The reason why we single out helium and why we discuss it in a book on cosmology is that there is now good reason to believe that the helium may have been formed in the early dense stages of the expanding Universe, whereas the heavier elements may have been formed in stars or massive objects after the development of galaxies.

The problem arises in the first place because abundance measurements have shown that roughly speaking 92 per cent of the atoms in our Galaxy are hydrogen, 8 per cent are helium while only one atom in a thousand is a heavier element. This striking disparity between the abundances of hydrogen and the other elements suggests the hypothesis that matter started off as hydrogen – the simplest of the elements – and that nuclear reactions occurred that led to the transformation of a small fraction of this material into heavier elements. On this view the Earth itself, which has lost most of its hydrogen by escape from its weak gravitational field, is a mere impurity speck.

There are two reasons for believing that the helium and the heavier elements had a different origin, and these two reasons fit well together. The first is that, as we shall see later, the stars in our Galaxy could have manufactured only about 10 per cent of the observed helium, whereas they could probably have manufactured all the heavier elements. Secondly, in the early dense stages of the Universe the helium could have been manufactured, but not the heavier elements.

THE ABUNDANCE OF HELIUM

Three methods have so far been developed for estimating the He/H ratio for various astronomical objects:

(i) The spectroscopic method which entails measuring the strengths of helium lines in the spectra of the objects concerned.

(ii) The cosmic ray method which works only for the Sun and involves measuring the relative abundances of various elements, including α-particles, in the solar component of the cosmic radiation.

(iii) The stellar structure method in which values for the helium content of stars can be deduced by fitting their observed properties to the predictions of stellar structure theory.

We consider each of these methods in turn.

(i) *The spectroscopic method.* The idea here is to measure the strengths of the helium lines in the spectra of various objects and then from the theory of the formation of spectral lines to deduce the relative abundance of the helium. For stars this procedure meets with several difficulties. It is not easy to observe the helium lines; when they are observed the theoretical path to an abundance is long, complicated and uncertain; when an abundance has been obtained it is not clear whether it refers just to the surface layers where the lines are formed or whether it is representative of the star's initial composition throughout.

The helium lines are hard to observe because the most prominent lie in the ultra-violet and so cannot be observed at the Earth's surface. Even ultra-violet observations from above the atmosphere are severely hindered by absorption due to interstellar hydrogen. To obtain strong lines in the visible the helium must be highly excited and so must be at the surface of a very hot star. According to our understanding of stellar evolution this means that the star must be either young or highly evolved. In neither case is there any guarantee that we are dealing with the initial helium composition of the Galàxy, uncontaminated by subsequent nuclear processes in the stars.

Helium lines have in fact been observed in various types of

object: main sequence O and B stars, highly evolved B stars, gaseous nebulae and planetary nebulae. Until recently these observations indicated that wherever helium could be observed at all it had a minimum abundance of about one helium atom to every eleven hydrogen atoms. Since, as we shall see, this is a very high helium abundance to have been produced by stars in the Galaxy, it is tempting to suggest that the minimum ratio of 1/11 represents the relative helium abundance when the Galaxy was formed, and that localised increases in the observed ratio have to do with the local production of helium after the Galaxy was formed.

Unfortunately for this simple picture, from 1966 onwards several observers have reported the existence of stars in which helium is underabundant, in some cases by as much as a factor of a hundred, as compared with the 'minimum' ratio of 1/11. There have been a number of attempts to explain these observations away by invoking special conditions at the surfaces of these particular stars, but these attempts have their own difficulties, and the subject is in an extremely obscure state. We prefer to avoid discussing this question further since such a discussion would involve complex details of the behaviour of stellar atmospheres which would take us a long way from the Universe, and in any case are not well understood. Even as we write observational evidence is beginning to accumulate that the data on the underabundance of helium may be misleading. In the hope therefore that some adequate explanation will be found, we shall retain the provisional hypothesis that the minimum ratio is, after all, close to 1/11.

(ii) *The cosmic ray method.* The helium lines in the spectrum of the Sun (which for this purpose is a cool star) are too weak to be used with any accuracy for an abundance determination. Instead we can exploit the fact that some time after an intense solar flare a burst of low energy cosmic rays which is known to be of solar origin arrives at the Earth. The abundance of α-particles relative to the common isotopes of carbon, nitrogen and oxygen in these cosmic rays probably reflects their relative abundance at the Sun's surface, since they all have the same charge to mass ratio and so are presumably accelerated in the same way. Observation confirms this argument for carbon, nitrogen and oxygen, and if extended to helium again gives for the He/H ratio the value 1/11 (or somewhat less according to a recent re-determination by D. L. Lambert).

(iii) *The stellar structure method.* Our understanding of stellar structure is now so advanced that we can tell a good deal about the composition of a star from its absolute luminosity and its surface temperature. In particular if we know its heavy element content (which can often be readily estimated from its spectrum) we can then infer its helium content. The advantage of this method is that the helium content refers to the bulk of the star and not just to its surface layers where the spectral lines are formed. The helium content of the Sun has been determined in this way with results in good agreement with the cosmic ray method.

The stellar structure method is particularly powerful if one studies not just one star at a time but a collection of stars in a cluster that are probably of the same age and have a well-defined main sequence. The method is not foolproof, however, since even the determination of a star's absolute luminosity and surface temperature has its difficulties. In fact the results so far obtained are consistent with most of the spectroscopic determinations in that they give helium abundances which are never less than 1/11 (even for some of the spectroscopically anomalous stars). Unfortunately the clusters concerned have formed fairly recently from interstellar material, so this does not tell us directly the initial helium content of the Galaxy.

Two variants of this method have recently been devised. One concerns the way in which old stars evolve off the main sequence as their hydrogen fuel is used up. Faulkner and I. Iben have studied this question, and again they find that they can explain the observations only if the He/H ratio is about 1/11. A similar result is obtained by R. Christy in his difficult non-linear calculations of the pulsations of variable stars. It appears that ionisation effects in the helium in such stars help to trigger off the pulsations, and unless the He/H ratio is around 1/11 the calculated light curves differ very appreciably from the observed ones.

There is no doubt that all these methods are highly sophisticated, and there is room for uncertainty in many stages of the argument. Nevertheless a very substantial modification would have to be introduced in each case if a model with a much lower helium content than 1/11 were for some other reason considered appropriate. We therefore, again, although still tentatively, conclude that we must find an explanation for a present He/H ratio in the Galaxy of at least 1/11.

HELIUM PRODUCTION IN STARS

The first question to ask is: how much helium has been formed by stars in the Galaxy during its lifetime of about 10^{10} years? We know that most stars generate their energy by converting hydrogen into helium in their deep interiors. Some of this helium will be transferred to interstellar space by explosive processes of various types, and so would contaminate the stars formed relatively recently from the interstellar medium. Can we not explain all the helium abundance determinations by appealing to processes of this type?

The answer appears to be 'no', by a factor of about 10. The luminosity of the Galaxy is about 10^{44} erg s^{-1}, and if this has been constant for 10^{10} years then altogether about 3×10^{61} ergs would have been produced. If we make the favourable assumption that all this energy has come from the conversion of hydrogen into helium, then about 10^{43} g of helium would have been produced (the binding energy of a helium nucleus being 2.5×10^{-5} ergs). However the total mass of the Galaxy is about 4×10^{44} g. Thus according to this argument the He/H ratio would be only 1/40 by mass or 1/160 by number. This is less than a tenth of the observed ratio of 1/11 by number.

This argument depends critically on the assumption that the luminosity of the Galaxy was not much greater in the past than it is now. However, some astronomers have suggested that when the Galaxy was very young there was a powerful burst of star formation, and that those stars were sufficiently massive to evolve rapidly and so to produce the necessary helium. When examined in detail this suggestion meets severe difficulties. The main problem is to ensure that the helium is transferred to interstellar space before it too is burnt in the massive stars and converted into heavier elements. Detailed calculations strongly suggest that the helium is in fact burnt before it is removed from the stars.

These calculations assume that hydrogen is not being continually fed into the central regions of the stars from their envelopes by convection. This absence of convection is normally regarded as being well-established, but recently P. Ledoux has questioned it, and has shown that substantial helium production would certainly occur if the stars remain well-mixed. Nevertheless at the present moment it seems more likely that the mixing was insufficient for this purpose.

Sufficient convection would be much more likely to occur in 'supermassive' stars, that is stars of 10^4 or more solar masses, rather than in the stars of a few solar masses that we have been considering so far. If such supermassive stars formed early in the history of the Galaxy they might have manufactured the necessary helium and blown themselves to bits by now. This remains a serious possibility, but how probable it is is hard to estimate. We still understand so little about the formation of the stars that we know to exist, that we can say nothing definite about the existence of other types of star. In recent years, at any rate, astronomers have come to realise how little they understand of the macroscopic forms that matter can take. The radio galaxies, QSOs, X-ray sources, infra-red stars and pulsars were all too exotic for their existence to have been predicted before they were observed. It will be a long time before we can safely pronounce on the formation of supermassive stars.

CONCLUSIONS

Having ended this chapter on such a note of uncertainty we feel obliged to try to sum up this complicated discussion. The reader will have noticed that as distinct from some of the other observations of cosmological interest, the measurement and interpretation of helium abundances draws on a wide range of extremely intricate and ill-understood astrophysical processes. The relevant arguments depend critically on certain details of stellar structure, stellar evolution, stellar atmospheres and even stellar formation, while none of these details are definitely established. This is a typical situation in astrophysics. One has to pick one's way through a complicated maze of uncertain considerations, each new result altering the delicate balance that has been achieved at any time. By great good fortune there is one new result that has altered the balance in favour of the minimum He/H ratio of 1/11 and of the cosmological origin of this ratio. This is the discovery of the cosmic microwave radiation. To appreciate the significance of this discovery we must first consider the processes that may have occurred in the early dense stages of the Universe. This question is discussed in detail in the next chapter.

12 The hot big bang

INTRODUCTION

The idea that the early dense stages of the Universe were hot enough to enable thermonuclear reactions to occur was first proposed by George Gamow in 1946. He hoped at that time that it would be possible to account for the observed abundances of helium and perhaps even of all the elements in terms of these reactions. This point of view has come to be known as the α-β-γ theory (after a key paper by R. A. Alpher, H. A. Bethe and Gamow himself) or as the hot big bang theory. It is now almost certain that this theory cannot account for the existence of the elements heavier than helium (with the possible exception of lithium-7), but that it can account for a helium abundance of one atom in twelve, the abundance required by most of the observations. In this way the behaviour of the Universe in its first few minutes may have left its permanent mark.

Perhaps the most exciting feature of this situation is that the heat needed to promote the formation of helium may also have left its own permanent mark on the Universe, namely, in the form of a universal radiation field that would now have a low but non-zero intensity. In fact by a curious perversity that we shall explain later, the *lower* the present intensity of this radiation field, and so the harder to detect, the *greater* the resulting abundance of helium. Thus if a sceptical physicist tries to throw cosmology out through the front door it comes in again at the back window.

In fact a universal radiation field has now almost certainly been detected, and its spectrum and intensity fit very well with the expectations of the α-β-γ theory (chapter 13). We must therefore take this theory seriously. To understand it we must understand the behaviour of a radiation field in the expanding Universe. This behaviour leads to the idea of a radiation-dominated Universe, whose properties we discuss in this chapter; the resulting nuclear reactions are left to the next chapter. We shall see in particular that the radiation would have rapidly reached thermodynamic equilibrium, and so be characterised simply by a temperature. This temperature then decreases as the Universe expands. The thermodynamic properties of such 'black body' radiation play an essential role in the discussion. The reader unfamiliar with these properties is asked simply to take on trust the equations that follow.

RADIATION IN AN EXPANDING UNIVERSE

It is important to compare the behaviour of matter and radiation in an expanding Universe. The density of material particles ρ_{mat}, and so the rest-energy density $\rho_{mat}c^2$, of matter decreases as the Universe expands in proportion as a volume element increases (we are here assuming that matter is conserved). In terms of the scale factor $R(t)$ of chapter 8 we have

$$\rho_{mat}c^2 \propto \frac{1}{R^3(t)}. \tag{1}$$

Now suppose that the Universe is filled with a homogeneous and isotropic distribution of radiation. This radiation, of course, moves with the speed of light relative to the matter, and with this motion is associated a red shift that reduces the energy of the radiation by the further factor $1/R(t)$ (p. 118). It is helpful here to think in terms of photons. The number of photons per unit volume decreases in the same way as the number of material particles per unit volume ($\propto 1/R^3(t)$), but whereas the rest-energy of a material particle remains the same, the energy of each photon decreases by the factor $1/R(t)$. We therefore have for the energy density $\rho_{rad}c^2$ of the radiation

$$\rho_{rad}c^2 \propto \frac{1}{R^4(t)}. \tag{2}$$

Perhaps an easier way of obtaining this important result is to use imaginary mirrors. Consider a small volume element of the Universe. Every time a photon leaves this element, on average a similar photon enters at the same place (we are assuming that the photons are distributed uniformly in all directions). The situation would be essentially unchanged, therefore, if we surrounded the volume element by perfectly reflecting walls that move out as the Universe expands. The red shift now arises from the Doppler effect associated with reflection at a moving mirror. The advantage of this point of view is that we have removed the complicated, and here irrelevant, cosmological aspects of the situation (the radiation at a given place having originated at a great distance, and so on). The behaviour of radiation in an expanding perfectly reflecting enclosure is well known, and is used in the standard elementary discussion of the thermodynamics of radiation (such as the derivation of Wien's law). To obtain (2) we simply use the adiabatic relation

$$pV^\gamma = \text{constant},$$

where p is the radiation pressure, V the volume and γ the ratio of the specific heats (which for isotropic radiation is $4/3$). Since $V \propto R^3(t)$ we have $p \propto 1/R^4(t)$. But for isotropic radiation the pressure is related to the energy density by the equation $p = \frac{1}{3}\rho_{\text{rad}}c^2$, so we again have

$$\rho_{\text{rad}} \propto \frac{1}{R^4(t)}.$$

Now if we compare (1) and (2) we see that if there is any radiation at all, then for sufficiently small $R(t)$ the energy density of radiation exceeds the rest-energy density of matter. In the point source models considered in this book $R(t)$ becomes arbitrarily small in the past and so we conclude, with Gamow, that the earliest phases of these models were radiation-dominated. How long this stage lasts we shall see presently. For the moment we simply recall the corresponding behaviour of $R(t)$ (chapter 8):

$$R(t) \propto t^{\frac{1}{2}} \quad (t \text{ small}). \tag{3}$$

We are now in a position to see whether thermal equilibrium would have been set up as a result of interactions between the radiation and the matter. Since the degree of excitation of the matter was very high at sufficiently early stages (it was formally infinite at $t = 0$) we can suppose the matter to be ionised and it will suffice for our purpose to consider just the inelastic (free-free) scattering of photons by electrons. If at any stage the scattering time was very much less than the time for the matter in the Universe to, say, halve its density, then we can be sure that thermal equilibrium would have been set up. It is a straightforward matter to show that this condition was easily satisfied at sufficiently early times.

Thus as long as there was some matter present to provide the necessary interactions we can be sure that the radiation in the early, radiation-dominated, stages of the Universe had an equilibrium (black body) spectrum completely characterised by a temperature T_{rad}. This temperature is related to the energy density ρ_{rad} by the usual black body formula

$$\rho_{rad} = aT^4_{rad},$$

where a, the Stefan constant, has the value 7.59×10^{-15} erg cm^{-3} deg^{-4}. Accordingly $T_{rad} \propto \rho^{\frac{1}{4}}_{rad} \propto 1/R(t) \propto 1/t^{\frac{1}{2}}$. General relativity gives the precise relation

$$T_{rad} = \frac{1.5 \times 10^{10}}{t^{\frac{1}{2}}_{sec}} \,^\circ K \quad (t \text{ small}). \tag{4}$$

Thus if radiation still dominated at a time of 1 second from the big bang, the radiation temperature would then have been 1.5×10^{10} °K. Actually such a temperature corresponds to an energy of about 1 MeV and so electron–positron pairs would have been created and would also have been in thermal equilibrium. When allowance is made for these and also for the presence of pairs of particles called electron-type and muon-type neutrinos, the total energy becomes $\frac{9}{2}aT^4$ (not $4aT^4$ despite the presence of the 4 relativistic fields: photon, electron pair, and two types of neutrino pair, because the electrons and neutrinos obey Fermi–Dirac statistics) and the temperature becomes

$$T_{rad} = \frac{10^{10}}{t^{\frac{1}{2}}_{sec}} \,^\circ K \quad (t \text{ small}),$$

a very simple result.

We must now consider what happens to the black body radiation at later times when it ceases to dominate. If there is no further interaction with matter our mirror argument assures us that the radiation would have retained its black body character, with the temperature obeying the adiabatic relation

$$T_{rad}V^{\gamma-1} = \text{constant},$$

which tells us that $$T_{rad} \propto \frac{1}{R(t)},$$ (5)

since $\gamma = \frac{4}{3}$ and $V \propto R^3(t)$. This is in agreement with (4) for the radiation-dominated phase, but the result (5) is quite general (in the absence of appreciable interaction with matter). By contrast, the temperature of the matter (in the absence of appreciable interaction with the radiation) satisfies the adiabatic relation

$$T_{mat}V^{\gamma-1} = \text{constant},$$

where now γ for the matter has the approximate value 5/3, that characterises a perfect gas. Thus

$$T_{mat} \propto \frac{1}{R^2(t)}.$$

We now have the following situation:

(i) The energy density of radiation decreases *more* rapidly with time than the rest-energy density of matter.

(ii) The temperature of radiation decreases *less* rapidly with time than the temperature of matter,* unless the radiation and matter are strongly coupled, in which case of course they have equal temperatures. Notice that the energy density and the temperature of black body radiation determine each other, while the rest-energy density and the temperature of matter are quite separate properties.

We now turn to the important question of how to specify in a convenient way the relative amounts of matter and radiation in the Universe. It is not convenient to compare their energy densities because such a comparison would be time dependent. It would be preferable to have a measure that is more or less independent of time. To find such a measure we recall that the energy density of

---

* We have used this result in chapter 10.

radiation falls off faster than that of matter because each photon gets red shifted as time goes on. Clearly then what we need to do is to compare the number density of photons with the number density of material particles; each decreases like $1/R^3(t)$, that is like T_{rad}^3. This last relation is a standard one for photons with a black body spectrum. Most of the energy is carried by photons whose individual energies are proportional to T_{rad}, and since the total energy density is proportional to T_{rad}^4 the number density of photons is proportional to T_{rad}^3.

Another way of making this comparison is to work in terms of the entropy density S of the black body radiation. Those unfamiliar with this notion may continue to think in terms of the number density of photons, which in fact is proportional to the entropy density of black body radiation. Now the thermodynamic theory of black body radiation tells us that

$$S = \tfrac{4}{3} a T_{\text{rad}}^3.$$

But as we have seen the concentration n of material particles is related to T_{rad} in the same way:

$$n \propto T_{\text{rad}}^3.$$

Hence the entropy per particle S/n $(= s)$ is independent of T_{rad} and so is independent of time. We shall therefore take the quantity s as our measure of the relative amounts of radiation and matter. This has the further advantage that it suggests to us that somewhere, somehow, irreversible processes occurred that produced this much entropy per particle. This certainly seems a more fruitful point of view than to suppose that the actual value of s in our Universe is one of its initial conditions, though some cosmologists do in fact suppose this.

To end this chapter, we shall determine the point at which the Universe ceases to be radiation-dominated. This moment will occur at later times the greater is the radiation–matter ratio, that is, the greater is the value of s. Of course there is no precise definition of radiation-domination, so we shall determine the time when radiation and matter had equal energy densities. In terms of grammes per cm^3 we have

$$\rho_{rad} = 8.42 \times 10^{-36} T_{rad}^4,$$

$$\rho_{mat} = \frac{4}{3} \frac{am_H}{s} T_{rad}^3,$$

$$= \frac{1.3 \times 10^{-38}}{s} T_{rad}^3,$$

where m_H is the mass of a hydrogen atom in grammes. These densities are equal when

$$T_{rad} = \frac{1.6 \times 10^{-3}}{s}.$$

Now so long as t is not too large we have

$$t_{sec} \sim \frac{10^{20}}{T_{rad}^2},$$

and so radiation and matter have equal energy densities at a time t_0 given by

$$t_0 \sim 3 \times 10^{25} s^2 \text{ seconds}.$$

The common density is then

$$\frac{6 \times 10^{-47}}{s^4} \text{ g cm}^{-3}.$$

The present matter density probably lies between $\sim 2 \times 10^{-31}$ g cm^{-3} (low density universe) and $\sim 2 \times 10^{-29}$ g cm^{-3} (high density universe). Since

$$\rho_{mat} \propto (1+z)^3,$$

the red shift z relative to us now of the equi-density moment is given in a low density universe by

$$1+z \sim \frac{7 \times 10^{-6}}{s^{\frac{4}{3}}},$$

and in a high density universe by

$$1+z \sim \frac{1.4 \times 10^{-6}}{s^{\frac{4}{3}}}.$$

To evaluate all these quantities we need to know s. This we can estimate in two ways, namely, by requiring that the helium formation in the radiation-dominated phase should give us the He/H ratio of $1/11$ and from measurements of the present temperature of the black body radiation. We shall see that these two methods give similar values for s, which encourages us to believe that cosmology and astrophysics may be closely linked together.

13 Helium formation in the hot big bang

INTRODUCTION

The first detailed calculation of helium formation in the hot big bang was made by Alpher, Bethe and Gamow in 1948. We now know that this calculation was wrong in an important aspect, but many of its ideas are valuable and we shall give a brief account of it here. Then we will consider the correct calculations, which have now attained considerable precision. These results permit us to place important restrictions on the fundamental quantity s, the entropy per particle, which we introduced in the last chapter. This quantity turns out to be in a sense very large, which raises the important and difficult question of discovering the irreversible processes that led to the production of so much entropy per particle. We end this chapter on a speculative note, by discussing possible mechanisms that might suppress the formation of primordial helium, while remaining within the limits of the hot big bang theory. At present it seems unlikely that these mechanisms operated but they are mentioned as a warning that the hot big bang theory may not be as simple in practice as its basic ideas are simple in principle.

THE α-β-γ THEORY

The basic assumption of this theory, apart from the hot big bang itself, is that initially matter was composed of neutrons. This is the very point that we must challenge later, but let us for the moment accept it. The relation between temperature (in °K) and time (in seconds) in the early stages would then have been

$$T = \frac{1.5 \times 10^{10}}{t^{\frac{1}{2}}}. \tag{1}$$

Now the half-life of a neutron for decay into a proton, an electron and an anti-neutrino is 700 seconds. That is to say, in the absence of further nuclear reactions there would have been an equal number of neutrons and protons at $t = 700$, at which time the temperature would have dropped to about 5×10^8 °K. However, if the matter density were high enough nuclear reactions would in fact have occurred at a rate faster than or comparable to $\frac{1}{700}$ s^{-1}. The slowest of these reactions is the one in which a neutron and proton come together to form a deuteron, the binding energy of the deuteron being emitted in the form of a γ-ray:

$$n+p \longrightarrow D+\gamma. \tag{2}$$

This reaction is rapidly followed by a series, of which the end product is mainly helium. Typical examples of such reactions are:

$$D+D\left\langle\begin{array}{l} {}^3\text{He}+\text{n} \\ {}^3\text{H}+\text{p} \end{array}\right.,$$

$${}^3\text{He}+\text{n} \longrightarrow {}^3\text{H}+\text{p},$$

$${}^3\text{H}+\text{D} \longrightarrow {}^3\text{He}+\text{n}.$$

All these reactions are much faster than (2), so the amount of helium formed depends essentially on how many deuterons are formed by (2). If the rate of this reaction at $t = 700$ s exceeds about $\frac{1}{700}$ s^{-1}, the protons combine with neutrons soon after they are produced by neutron decay, and nearly all the material is turned into helium. If the rate is less than $\frac{1}{700}$ s^{-1}, then owing to the expansion of the Universe, the material density drops before much deuterium is formed, and this reduces the rate of reaction (2) still further. Since this rate depends on the product of the neutron density and the proton density, and on the temperature, it decreases very much faster than the inverse time-scale of the expansion, and essentially no helium is formed. The transition from no helium to all helium thus occurs over a small range of the parameter s ($\propto T^3/n$) which specifies the density of matter at $t = 700$ s.

It follows that the correct value of s would be closely specified by the requirement that the amount of helium formed should be roughly comparable with the amount of hydrogen left. This value of s is about 10^{-3}. For thermodynamic reasons it is more convenient to express s in terms of the Boltzmann constant k ($= 1.37 \times 10^{-16}$ erg deg^{-1}). We then have for the required value of s:

$$s \sim 10^{13}k. \qquad (3)$$

What is remarkable about this result is that we can confidently reject it as being larger than the observations permit. As we saw at the end of the last chapter s is independent of time, and so its present value would also have to be $10^{13}k$ on this theory. Moreover, if we allowed for the local production of entropy by irreversible processes in galaxies, then the present value of s would be higher still. However, there is a sufficient disagreement with observation if we consider only the black body radiation referred to in (3).

Now the smallest possible value of the present particle density is about 10^{-7} cm^{-3}. Thus the smallest present value for the black body temperature which would yield $10^{13}k$ for s is given by

$$10^{13}k = \frac{\frac{4}{3}aT_{\text{now}}^3}{10^{-7}},$$

from which $T_{\text{now}} \sim 25\ ^\circ\text{K}.$

This is much too high to be acceptable. Quite apart from the question of whether the observed microwave background radiation discussed in the next chapter actually has a black body spectrum, those microwave observations clearly rule out a black body radiation field hotter than 3 $^\circ$K. Similar remarks apply to the behaviour of interstellar cyanogen which, as is also explained in the next chapter, serves as a thermometer for the radiation field at a wavelength of 2.6 mm. Whatever the correct interpretation of the cyanogen observations, they again give an upper limit of 3 $^\circ$K for a black body component. Finally, the effect of a 25 $^\circ$K radiation field on cosmic ray protons and electrons would easily have been noticed as discussed in chapter 15. This last argument is not quite so direct as the other two and does not lead to quite such a firm upper limit. But it is very unlikely that a black body temperature much in excess of 10 $^\circ$K could be tolerated on cosmic ray grounds.

The excess prediction of T_{now} by a factor ~ 10 is very serious because s depends on the third power of T. The highest value of s

we can tolerate is thus about $10^{10}k$. Moreover, if the present particle density were now about 10^{-5} cm^{-3}, s_{max} would be about $10^8 k$. In either case the α-β-γ theory would predict that essentially all matter must be in the form of helium, in disagreement with observation.

HAYASHI'S THEORY

The fallacy in the α-β-γ argument was pointed out by Hayashi in 1950. At times earlier than 2 seconds the temperature would have exceeded 10^{10} °K, and this is beyond the threshold for the creation of electron-positron pairs. (One pair requires $2m_0c^2$ of energy, which is about 1 MeV; at 10^{10} °K many photons have this much energy.) The effect of these pairs is twofold. The first, relatively minor effect, is that their gravitational field changes the time-scale of the expansion. If we include the neutrino pairs that would also be excited, the relation (1) becomes

$$ T = \frac{10^{10}}{t^{\frac{1}{2}}}, $$

a result that we mentioned in the last chapter.

A much more important effect of the electron pairs is the following. Instead of half the neutrons taking 700 seconds to turn into protons, they would take a very much shorter time. The reason is that the neutrons can interact with the thermally excited positrons, through the so-called weak interactions, to produce protons and anti-neutrinos:

$$ n + e^+ \longrightarrow p + \bar{\nu}. $$

The reaction time for this process turns out to be less than the expansion time for $t < \sim 1$ second, that is, for $T > 10^{10}$ °K. Since the reverse process also occurs rapidly through the presence of the thermally excited anti-neutrinos, Hayashi proposed that at temperatures above 10^{10} °K *there was complete thermal equilibrium between all forms of matter and radiation*. Below 10^{10} °K the weak interactions can no longer maintain the neutrons in statistical balance with the protons because the concentration of electron pairs is beginning to drop abruptly. The n/p ratio is then 'frozen in', until a few hundred seconds have passed and neutron decay begins to be appreciable. This frozen-in ratio, corresponding to thermal equilibrium at a temperature somewhat below 10^{10} °K, is about 15 per cent.

This change in the n/p ratio at $t \sim 100$ seconds from that assumed in the α-β-γ theory completely alters the final helium abundances. Of course, for sufficiently low material densities the key reaction (2) never has a chance to get going and not much helium is formed, as in the α-β-γ theory. At higher densities the reaction does get going, but at temperatures greater than 10^9 °K there are enough photons to disintegrate the deuterons so rapidly that it is still ineffective. This is no longer true at 10^9 °K ($t = 100$ seconds), so this is when the helium gets built up. At this stage the neutrons have their frozen-in abundance, and all of them combine with protons to form helium.

We now come to the important difference from the α-β-γ case. The frozen-in abundance of neutrons hardly depends on the material density, that is, on s, but does depend on the temperature and the properties of the weak interactions. Thus so long as the density is great enough for the key reaction (2) to be more rapid than the expansion time, this fixed concentration of neutrons is incorporated into helium nuclei *however great the material density is*. Thus the dependence of helium abundance on s has a plateau, which in fact is only appreciably departed from when s is so low that the Universe is still matter-dominated at 10^{10} °K. We would then be back to the α-β-γ case and most of the matter would be turned into helium.

Detailed calculations show that the plateau continues for a range of about six powers of 10 in s. Even without detailed calculation we can estimate the height of the plateau from our estimate of 15 per cent for the frozen-in ratio of neutrons. Since there are as many protons as neutrons in a helium nucleus, and since on the plateau all neutrons are incorporated into helium nuclei, the final helium abundance on the plateau should be 30 per cent by mass. This is essentially the same as the 27 per cent we need according to chapter 11.

A determination of the exact value of the plateau abundance and the (small) dependence on s in this range requires accurate knowledge of the weak interactions and a numerical calculation of all the reactions between the light particles, only a few of which were listed earlier. Hayashi's original value for the plateau abundance was 40 per cent, which is rather higher than the observations permit. A later series of calculations by Alpher, J. W. Follin and R. C. Herman, Hoyle and R. J. Tayler, Y. N. Smirnov, and P. J. Peebles gave about 35 per cent, which is still rather high. If we descend below the plateau to avoid this difficulty, we run into the same problem as the α-β-γ-theory, an unacceptably high value for the present-day black body temperature.

This was the situation until late in 1966 when Peebles published another calculation in which he improved the accuracy of his numerical integrations. He then obtained 27 per cent for the plateau abundance, in very good agreement with what is required. His results were confirmed in 1967 by R. V. Wagoner, Fowler and Hoyle who made a much more elaborate calculation. This success of the theory is so encouraging and the problem is so important, that it is worth looking at these results and their implications in more detail.

IMPLICATIONS OF THE WAGONER–FOWLER–HOYLE
CALCULATIONS

These calculations took into account all the reactions that can occur between the light elements with the most up to date values for their cross-sections. The build-up of heavier elements was also considered (144 different reactions being included) with results that will be described later. The final helium abundance as a function of s is plotted in fig. 47. The region above the plateau, where the Universe is matter-dominated during the nuclear reactions, is not included because the computer program would have required considerable modification to handle this regime.

The plateau begins at $s \sim 10^4 k$ with a helium abundance of 40 per cent and ends at $s \sim 2.5 \times 10^{10} \, k$ at a level of 24 per cent. The material density at a given temperature can thus change by a factor 2.5×10^6 for a change in the helium abundance of less than a factor 2. The observed helium abundance is thus not a sensitive indicator of the value of s. Perhaps it is reasonable to demand that the calculated helium abundance X_{He} should lie between 25 and 30 per cent. In that case we have

$$3 \times 10^7 k \leqslant s \leqslant 10^{10} k \quad (30\% \geqslant X_{He} \geqslant 25\%).$$

We can sharpen this inequality if we recall the limits on s imposed in the first section of this chapter. For a low density universe there is no further restriction and we have

$$3 \times 10^7 k \leqslant s \leqslant 10^{10} k \quad (n \sim 10^{-7} \text{ cm}^{-3}),$$

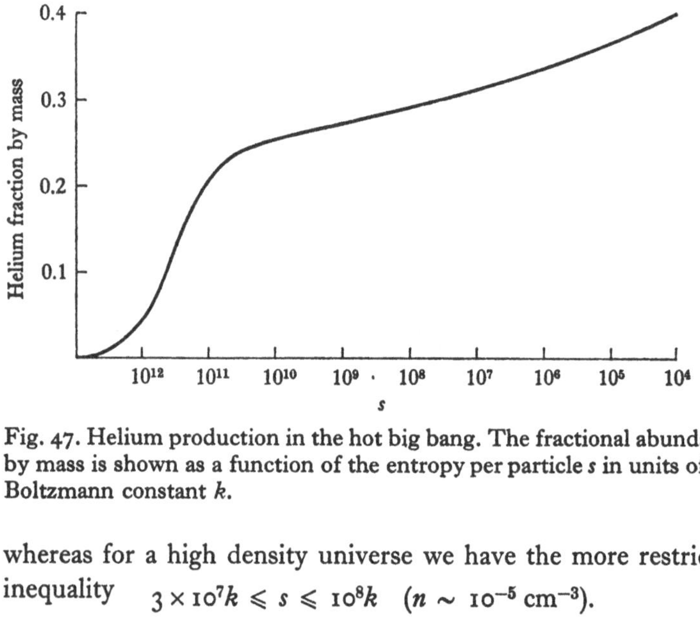

Fig. 47. Helium production in the hot big bang. The fractional abundance by mass is shown as a function of the entropy per particle s in units of the Boltzmann constant k.

whereas for a high density universe we have the more restricted inequality $\quad 3 \times 10^7 k \leqslant s \leqslant 10^8 k \quad (n \sim 10^{-5} \text{ cm}^{-3}).$

If we accept that the present temperature of the black body radiation is 3 °K then we have the two extreme results

$$s = 10^{10} k, \quad X_{He} = 25\% \quad (n \sim 10^{-7} \text{ cm}^{-3}),$$
$$s = 10^8 k, \quad X_{He} = 29\% \quad (n \sim 10^{-5} \text{ cm}^{-3}).$$

According to Wagoner, Fowler and Hoyle their calculation of X_{He} is accurate to 1 per cent.[*] Thus if the observed minimum helium

---

[*] However, recent experimental evidence suggests that the half-life of the neutron is about 10 per cent less than was previously believed. The resulting increase in the weak interaction coupling constant would imply that neutrons remain in thermal equilibrium for a longer time, so that the frozen-in abundance of neutrons would be reduced. The final helium abundance would also be reduced – by about 10 per cent according to Tayler.

abundance could be obtained with an accuracy of, say, 10 per cent, one would obtain from this theory quite close limits on the present matter density n, which would be a result of great significance as we have seen in earlier parts of this book.

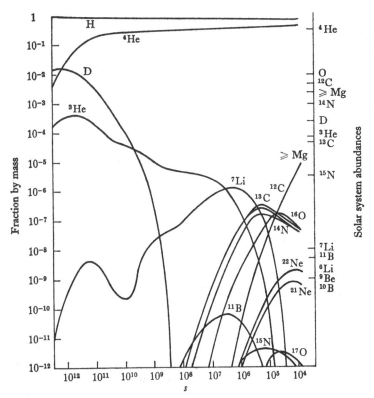

Fig. 48. Element production in the hot big bang, according to Wagoner, Fowler and Hoyle. As in fig. 47, the fractional abundance by mass is shown as a function of the entropy per particle. (After *Astrophys. J.* **148**, 21 (1967).)

We must also mention briefly the results of Wagoner, Fowler and Hoyle for the build-up of other elements. Their complete results are shown in fig. 48. It will be seen that for $s \sim 10^{10}k$ the D and ^{3}He abundance agree well with observation and that even Li can be brought into line if one admitted localised irregularities in the distribution of matter in which s were reduced by a factor $\sim$ 100. This consideration would support a low density universe. However, as Wagoner, Fowler and Hoyle point out, there are other non-cosmological mechanisms for making (and destroying) these light elements, and the present situation is quite unclear.

Fig. 48 also gives the results for the primeval build-up of the elements heavier than helium. The main difference from earlier calculations is that previously the absence of stable nuclei of atomic weights 5 and 8 completely prevented the build-up of heavier elements. Reactions are now known that bridge these gaps, but even so far too few heavy elements are formed. We must therefore still appeal to non-cosmological processes for the formation of these elements, processes such as nuclear reactions in hot stars or massive objects. This is not really surprising because there is now considerable evidence that the concentration of heavy elements in the Galaxy has changed with time. Old stars, for instance, tend to have a lower metal content than young stars. The correlation of element abundance, stellar type, and motion and position in the Galaxy has become an important branch of astrophysics which should lead to a better understanding of the formation of the heavier elements. Since this is not strictly speaking a cosmological problem we shall have to pass it by.

MECHANISMS FOR SUPPRESSING HELIUM FORMATION

We now consider the possibility that the low helium abundance of the old B stars is genuine, indicating a very low primeval abundance of helium. It does not seem possible to have $s > 10^{10}k$ because that would require there to be too much black body radiation now. If the Universe were irregular at the time of element formation there might be localised regions where s was substantially greater than $10^{10}k$, so that very little helium would be formed in those regions. However such regions would have a much lower material density than average, and it is perhaps unlikely that a galaxy could develop from such a region. Indeed this argument suggests the opposite conclusion, that the helium abundance in intergalactic space might be less than in the Galaxy. As we have seen (p. 136) it might be possible to examine this question by observations made from above the Earth's atmosphere. Several other mechanisms for suppressing the helium while retaining the hot big bang have been proposed. Some involve modifications of Einstein's general theory of relativity, which we shall not consider in this book. Two rather more conventional mechanisms should be mentioned here, although at the time of writing neither seems likely or attractive.

The first is the possibility that the Universe is filled with electron-type neutrinos or anti-neutrinos with about 10^{10} neutrinos per material particle. Such a neutrino flux could not be detected to-day, so that there is no empirical objection to introducing this hypothesis. At first sight it might seem that there should be no theoretical objection either, since we have already accepted the idea that there might be about 10^9 photons per material particle in the Universe. There is one important difference, however, in the two situations. Irreversible processes can occur that create entropy and this entropy can be in the form of photons; photon number is not conserved. By contrast, elementary particle physicists believe that there is a conservation law for the light particles or leptons (neutrinos, electrons, muons) which in the present context means that no processes occurring in the Universe can change the number of neutrinos per particle. This number is fixed once and for all. Admittedly we do not understand at all what determines this number, but most physicists would be reluctant to have it of the order of 10^{10} rather than of the order of unity (or zero), unless it were absolutely necessary.

We have still to explain why a preponderance of neutrinos would affect the helium production. The reason is that reactions of the type

$$\nu + n \longrightarrow p + e^-$$

would go very fast, converting most of the neutrons into protons. There would then be very few neutrons left to undergo the key reaction

$$n + p \longrightarrow D + \gamma$$

and very little helium would be formed.

Similarly if there were a preponderance of anti-neutrinos, the protons would be rapidly removed by the reaction

$$\bar{\nu} + p \longrightarrow n + e^+$$

and there would be few protons left to undergo the key reaction. Moreover in this case the neutrons cannot decay at first because the anti-neutrinos emitted in this decay are subject to the exclusion principle and cannot find an empty level to occupy (if the initial preponderance of anti-neutrinos is sufficiently high). Since the anti-neutrinos are diluted by the expansion of the Universe the time would come when the neutrons can decay, but by then the temperature would be too low for the key reaction to occur. Again the result is very little helium.

Another mechanism for suppressing the helium involves the possibility that the Universe in its early stages was not isotropic. This would have had the effect of decreasing the time-scale for the expansion of the Universe (in analogy with the fact that a two-dimensional collapse in Newtonian dynamics is more rapid than a three-dimensional collapse). There would then be less time available for the nuclear reactions to occur that lead to the build-up of helium. A convenient measure of the amount of anisotropy present can be based on the fact that it decreases rather rapidly as the Universe expands (in analogy with the fact that the anisotropy of a collapsing body in Newtonian dynamics increases as the collapse proceeds). The time during which the anisotropy is appreciable is then a useful measure of the initial amount of anisotropy. It has been calculated that, in order to suppress the helium, this time has to be at least 10 years in a low density Universe and 20 000 years in a high density Universe.

These calculations neglect the fact that there may be other processes besides the expansion of the Universe that tend to reduce any initial anisotropy. In particular the neutrinos present at temperatures just above 10^{10} °K couple the expansion rates in different directions and tend to equalise these rates, that is, to produce isotropy. This mechanism was pointed out by C. W. Misner and by Doroshkevich, Zel'dovich and Novikov in 1967. Misner's detailed calculations for a particular class of cosmological models and a wide range of initial anisotropies show that any anisotropy remaining at temperatures $\sim 10^9$ °K, when the nuclear reactions are important, would be far too small to influence the helium production. This important work is further discussed in chapter 16, where the isotropy of the Universe is examined in more detail. The calculations will have to be extended to other types of cosmological model before we can be sure that the helium has not been suppressed by anisotropies, but such a mechanism for the suppression has clearly lost much of its attractiveness. Moreover, the observed low helium abundances may well be misleading, in which case the hot big bang theory itself severely limits the aniso-tropy that can prevail at 10^9 °K. This is clearly a problem in which we are likely to see great development in the next few years.

14 The detection of cosmic microwave radiation

INTRODUCTION

It follows from our discussion in the last three chapters that the successful detection of cosmic black body radiation would be of the first importance. It would draw together the various ideas underlying the cosmological approach to the helium problem and the thermodynamic behaviour of radiation in an expanding universe. Moreover, as we shall see, its existence and degree of isotropy are of great importance in other problems too. Yet we have to record that its first detection was an accident. How did this come about?

There appear to be a number of reasons. First of all the original α-β-γ theory, which predicted a present black body temperature of about 25 °K, was formulated in 1948. At that time it would not have been possible to detect such a radiation field (although the suspicion that 25 °K was too high a temperature could have been formulated even then because of the cyanogen problem that we discuss at the end of this chapter). The essential problem in detecting any⋅ proposed radiation field is to estimate the wavelengths at which such a field would not be too much dominated by other sources of radiation. In the case of 25 °K black body radiation, optical wavelengths would be out of the question. The spectrum peaks at 100 microns (10^{-2} cm), that is, in the far infra-red (which itself is heavily absorbed by the Earth's atmosphere), and from there to the visible the spectrum cuts off more or less exponentially, and so is swamped by the ordinary background light of the night sky.

Nowadays we would ask whether the radiation could be detected at radio wavelengths, but in 1948 radio-astronomy was in its infancy. In fact at wavelengths longer than 100 cm the effective temperature of the background due to our own Galaxy is much greater than 25 °K. But between wavelengths of 50 cm and 1 cm (below which atmospheric absorption is important) the 25 °K radiation field would have dominated. The reason is that the galactic background has a non-thermal spectrum, its effective temperature dropping as the wavelength λ is lowered in proportion to $\lambda^{2.7}$. By contrast, the black body radiation field has a temperature that is independent of wavelength, so that at low enough wavelengths it would dominate the radiation from the Galaxy. Even black body radiation with a temperature as low as 1 °K would dominate below about 13 cm.

Unfortunately by the time radio-astronomy had developed to the point where this was clearly understood the attention of interested scientists had been diverted from the α-β-γ theory by its inability to account for the formation of elements heavier than helium, and by the partial success of stellar theories of nucleogenesis. Despite the fact that the helium problem remained unsolved by these theories, people forgot about Gamow's prediction that the Universe today should be filled with black body radiation. There matters stood until the summer of 1964 when R. H. Dicke of Princeton University independently conceived the same idea of a hot big bang. Dicke's point of departure was not a detailed calculation of primordial helium formation but rather the possibility that the present expansion of the Universe was preceded by a collapse in which high temperatures were generated. Dicke realised that he could test this idea by attempting to detect the relic black body radiation at a wavelength of a few centimetres where, if it existed, it would dominate over other sources. Accordingly in the fall of 1964 he and his colleagues started to build the necessary equipment. As luck would have it, before their first measurements were made it was announced by A. A. Penzias and R. W. Wilson of the Bell Telephone Laboratories, who were working near Princeton at Holmdel, New Jersey, that they had detected an unwanted and unexpectedly large background at

SMC

7 cm, whose effective temperature was about 3.5 °K. Dicke, together with his colleagues Peebles, P. G. Roll and D. T. Wilkinson, immediately proposed that this excess background was indeed what they had been planning to find, namely cosmic black body radiation from the hot big bang. To test this suggestion the intensity of the background at other wavelengths had to be determined, to see whether it has a black body spectrum. We must now consider the various measurements that have been made.

MICROWAVE MEASUREMENTS OF THE COSMIC
BACKGROUND

It is not easy to make an absolute measurement of the intensity of the cosmic background. Usually in radio-astronomy one is content to make relative measurements of one part of the sky with respect to another, or of one discrete source with respect to another. There is then no need to calibrate one's receiver in absolute terms. Here, however, it is the essence of the problem to obtain an absolute measurement. In addition one has to allow for emission by the atmosphere and the ground, for losses in the equipment, and for the fact that the signal one is looking for is likely to be thousands of times weaker than the receiver noise.

This latter difficulty exists in many radio-astronomical measurements, and is usually avoided by a device designed by Dicke many years earlier. The receiver is periodically switched between the antenna and a reference source, which for a low temperature measurement might be a resistor dipped in liquid helium. The receiver output would then contain a periodic component whose frequency is the switching frequency and whose amplitude is a measure of the temperature difference between antenna and reference source. This signal is far weaker than the receiver noise but since it has a well-defined frequency it can be separated out by means of an amplifier sharply tuned to this frequency.

The emission from the atmosphere is mainly due to molecular oxygen and water and can be measured by tilting the antenna into different directions that correspond to different path lengths through the atmosphere. Such measurements give results in good agreement with theory, and no particular problem is raised by them. The ground emission is more troublesome and it is best to avoid it as much as possible by using a horn-shaped antenna rather than the more usual paraboloid.

Now it so happened that Penzias and Wilson had been making radio-astronomical measurements with a horn antenna originally designed to receive signals reflected from the Echo satellites. They had been puzzled by what seemed to them excess noise in their instrument, and a careful study then showed that they were picking up a background that was isotropic to a precision of a few per cent and was about a hundred times more intense than they could account for in terms of known radio sources. This was the 3.5 °K measurement at 7 cm to which we have already referred. They later modified their horn feed and arrived at a final result of 3.1 ± 1 °K.

This measurement has been followed by many others. First came Roll and Wilkinson in 1966 who obtained a temperature of 3.0 ± 0.5 °K at a wavelength of 3 cm. Then T. F. Howell and Shakeshaft at Cambridge found 2.8 ± 0.6 °K at 21 cm. At this wavelength the Galaxy makes an appreciable contribution that was estimated, by extrapolating its spectrum, to be 0.5 ± 0.2 °K. This contribution has already been subtracted out to give the result of 2.8 °K for the extragalactic contribution.

The present situation is depicted in fig. 49. The observations fit a black body curve reasonably well, with a temperature of 2.7 ± 0.2 °K.* For convenience we shall still refer to it as the 3 °K background. It is evident from the figure that the most characteristic part of the black body spectrum is near its peak at about 1 mm, beyond which the intensity drops very rapidly as the wavelength decreases. The most critical test of the black body nature of the spectrum would be to follow it over the peak. Unfortunately observations cannot be made from the ground at wavelengths in the vicinity of 1 mm, because in this wavelength range the Earth's atmosphere is heavily absorbing. Several groups are planning to make observations from above the Earth's atmosphere, and two

---

* (*Note added in proof.*) The latest ground-based measurement at 3.3 mm corresponds to a temperature of 2.61 ± 0.25 °K. The best fit to all the observations is given by a temperature of 2.65 ± 0.09 °K.

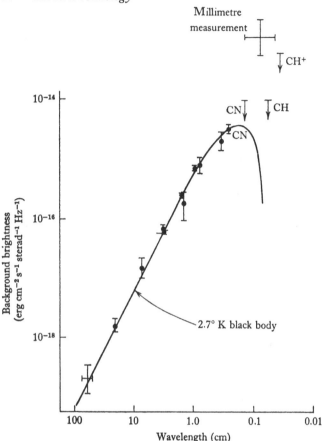

Fig. 49. Measurements of the isotropic background at radio wavelengths, compared with a 2·7 °K black body spectrum. The upper limits from interstellar molecules conflict with the millimetre measurement, which may refer to a discrete line superposed on the continuous background.

groups have already reported some successful results in the wavelength range 0.4–1.3 mm. The background intensity in this range appears to be surprisingly high and may have a local origin, e.g. in the upper atmosphere or interplanetary space. It is unlikely to represent the diffuse background in regions on a larger scale, for reasons explained in the next section. Nevertheless these first millimetre measurements present an interesting unsolved problem.

CYANOGEN MEASUREMENT OF THE COSMIC BACKGROUND

Fig. 49 and table 4 (see p. 183) also contain a value for the temperature of the background at 2.6 mm, an interesting wavelength since it is closer than the microwave measurements to the peak of the 3 °K black body spectrum at 1 mm, but nevertheless a wavelength that is heavily absorbed by the Earth's atmosphere. How this value was obtained is perhaps the most remarkable part of this remarkable story. It was in fact first obtained in 1941, but its significance was not recognised at that time. We may begin the story with a now famous quotation from the last page but one of Herzberg's standard book on molecular spectra published in 1941 (*Molecular Spectra and Molecular Structure. Volume 1: Spectra of Diatomic Molecules*): 'From the intensity ratio of the lines [of CN] with $K = 0$ and $K = 1$ a rotational temperature of 2.3 °K follows, which has of course only a very restricted meaning.'

Herzberg is referring to the observations made by A. McKellar of the Dominion Observatory of Canada, who found in the spectra of several stars absorption lines due to interstellar cyanogen, CN. When the CN molecule is non-rotating it has an absorption line in the violet at 3874.6 Å. However, if it is in its first rotational state this absorption line shifts to 3874.0 Å. For just one star, ζ Ophiuci, McKellar was able to measure both lines, and to estimate their relative strengths. This is a measure of the relative number of molecules in the ground state and the first rotational state. This relative number can be expressed in terms of an excitation temperature – if the molecules were in equilibrium in a heat bath at that temperature then they would possess the observed degree of excitation. Usually in interstellar space molecules are not thought of as being in contact with a heat bath – they are excited by collisions with other particles or with radiation that does not have a thermal spectrum. This is what Herzberg meant when he said that the excitation temperature of 2.3 °K had only a very restricted meaning.

Nevertheless it was realised in 1941 that the degree of excitation corresponding to 2.3 °K, namely, about a quarter of the molecules being in the upper rotational state, was very high if it was to be explained in terms of collisional excitation either by hydrogen atoms or by starlight. This anomaly remained an unsolved problem, though not a highly publicised one, for 25 years. Then in 1966, after the early microwave measurements of the cosmic background and the suggestion that this background has a black body spectrum at all wavelengths, it occurred independently to Field, to Shklovsky and to N. J. Woolf that the CN molecules may after all be in a heat bath, at a temperature of about 3 °K, which is close to McKellar's value of 2.3 °K. The wavelength of the radiation that can excite CN to its first rotational state is 2.6 mm, so the CN is acting as a thermometer for radiation of this wavelength, nicely filling in a part of the spectrum that cannot be observed below the atmosphere.

At this point a complication set in. G. Münch had suggested two years earlier that the interstellar CN may be in an environment influenced by ultra-violet radiation from the star whose spectrum is being studied. This environment would then contain an appreciable number of free protons and electrons as well as hydrogen atoms. This would be important because a proton or an electron is much more effective than a hydrogen atom in exciting CN by collisions. The reason for this is that the CN molecule has a permanent electric dipole moment which interacts much more strongly with a charged particle like a proton or an electron, than with an uncharged hydrogen atom. It turns out that a density of free protons of 1 cm^{-3} would suffice to account for the observed excitation. Such a density in the neighbourhood of ζ Ophiuci would be quite reasonable.

Fortunately there is a good way of distinguishing between the two excitation mechanisms. The degree of collisional excitation depends on the density of free protons, and this would be expected to vary considerably in different regions of the Galaxy. On the other hand the black body excitation should be universally the same. It is important therefore to determine the degree of excitation of the CN in the spectra of as many stars as possible located in different regions of the Galaxy. The results to date of this programme are as follows.

First of all the spectrum of ζ Ophiuci was re-observed by Field and J. L. Hitchcock (using plates taken by G. H. Herbig), and by P. Thaddeus and J. F. Clauser. These observers obtained excitation temperatures of 3.22 ± 0.15 °K and 3.75 ± 0.50 °K respectively.

Thaddeus and Clauser also found a temperature of about 2.7 °K from the absorption spectrum of another star ζ Persei which is in a different part of the sky from ζ Ophiuci. More recently Clauser has developed a technique for digitally summing the faint spectra on large numbers of star plates from the Mount Wilson Observatory. Eleven stars have been analysed in this way and all yield an excitation temperature of about 3 °K (table 4). Of course in any particular case there might be an appreciable contribution from proton or electron excitation, but the uniformity of the results suggests that in the regions so far observed this has not happened.

TABLE 4 *The rotational temperature T of the* CN (o, o) *violet band observed in absorption against eleven stars, corrected for the finite optical depth of the lines*

Uncertainties are estimated from the plate grain. V is the visual magnitude and Sp the spectral class of the stars. The heliocentric radial velocities of CN and H^I 21 cm are included in those instances where 21 cm peaks are found to be well defined.

Star	V	Sp	T(CN) (°K)	v(CN) (km s^{-1})	V(H^I) (km s^{-1})
ζ Oph	2.56	O9.5 V	2.74 ± 0.22	-14.8	-12.7
ζ Per	2.83	B1 Ib	2.82 ± 0.30	12.6	13.4
55 Cyg	4.83	B3 Ia	< 5.5	—	—
AE Aur	5.3	O9.5 V	3.5 ± 2.3	—	—
20 Aql	5.37	B3 IV	2.5 ± 1.8	-12.2	$-2.0, -11.5$
HD 12953[c]	5.68	A1 Ia	3.7 ± 0.7	—	—
13 Ceph[d]	5.79	B8 Ib	2.8 ± 0.4	—	—
HD 26571[a]	6.10	B8 II–III	~ 3		
χ Per	6.08	O	2.8 ± 0.8	13.5, 23.2	13.3, 22.8
BD+66° 1675[b]	9.05	O7	2.39 ± 0.4	-17.2	$-14.2, -19.5$
BD+66° 1674[b]	9.5	O	2.45 ± 0.6	-17.4	$-14.2, -19.5$

(a) G. Herbig, Lick 120-inch Coude, private communication.
(b) From spectra taken by G. Münch with the 200-inch telescope.
(c) M. Peimbert, 120-inch Coude, private communication.
(d) V. Bortolot and P. Thaddeus, 120-inch Coude.

Great interest also attaches to the possibility of observing absorption from the *second* rotational state of CN. Molecules in this state act as a thermometer for radiation of wavelength 1.3 mm, very close indeed to the peak of the black body spectrum. The negative observations of V. J. Bortolot, Thaddeus and Clauser enabled them to place an upper limit of 4.7 °K on the radiation temperature at 1.3 mm, and this upper limit is marked on fig. 49. They also attempted to observe absorption lines from rotational states of CH and CH+. Their failure to do so implied upper limits on the background temperature at 0.56 and 0.36 mm of 5.1 and 8.1 °K respectively.

These upper limits suggest that the intense millimetre background mentioned in the last section does not extend to the Galaxy as a whole, unless it is restricted to wavelength intervals that do not overlap the relatively narrow wavelength ranges involved in the molecular absorption lines. A further observational argument against the existence of an intense flux of millimetre photons in the Galaxy will be given in the next chapter.

CONCLUSIONS

So much significance attaches to the black body radiation if it is present, that it is necessary to scrutinise all these measurements, of microwave emission, as well as of CN absorption, with the greatest care. None of the evidence to date is beyond criticism, but taken together it must be admitted that it is very impressive. For the purpose of exposition in this book we shall assume that the black body radiation field exists, while recognising that more definite proof would be highly desirable. The decisive test will probably come from future observations in the millimetre range, where the spectrum would be expected to drop rapidly. Even here there may be complications, as we have seen, if there are unsuspected localised sources of millimetre radiation. In addition the black body radiation may interact appreciably with a re-heated intergalactic gas (p. 141), which could modify its spectrum and polarise it to a measurable extent. We may expect much research to be carried out on these problems in the near future.

15 Astrophysical effects of the cosmic microwave radiation

INTRODUCTION

From a laboratory point of view 3 °K is a very low temperature. Indeed to measure it the microwave observers had to use a reference terminal immersed in liquid helium. Nevertheless from an astrophysical point of view 3 °K is a very high temperature. A universal black body radiation field at this temperature would contribute an energy density everywhere of 1 eV cm^{-3}. As we saw in chapter 2 this is just the energy density *in our Galaxy* of the various modes of interstellar excitation – starlight, cosmic rays, magnetic fields and turbulent gas clouds. So even in our Galaxy the cosmological background radiation would be for many purposes as important as the well-known energy modes of local origin. In intergalactic space, these localised energy densities probably drop off by a factor between 100 and 1000, whereas the black body component would maintain its energy density at 1 eV cm^{-3}. It would thus become the dominant form of energy density in intergalactic space with the exception of the rest-energy density of matter itself. Even then, in a low density universe with $n \sim 10^{-7}$ cm^{-3} the energy density in the radiation field would be as much as 1 per cent of the rest-energy density in matter.

Under these circumstances we might expect the microwave radiation field to exert a noticeable influence on astrophysical processes, especially on those involving high energy particles or radiation that can interact with the microwave photons. This is the question we shall study in this chapter. While no known effect can definitely be attributed to this interaction, it is possible to make well-defined predictions that are amenable to future test. At the same time the absence of any clear effect to date puts an upper limit of about 10 °K on the temperature of a possible universal black body radiation field.

It will be useful for what follows to state the following properties (in round numbers) of a 3 °K radiation field. The number density of photons is 10^3 cm^{-3}, and since the total energy density is 1 eV cm^{-3} the mean energy per photon is 10^{-3} eV. These estimates are useful for making quick calculations of many of the effects of the radiation field, without having to consider in detail the full energy range of the photons in the black body spectrum.

RELATIVISTIC ELECTRONS AND THE MICROWAVE RADIATION FIELD

Relativistic electrons can collide with microwave photons and thereby transfer energy to them. In the astrophysical literature this process is called the inverse Compton effect. A typical scattered photon would have an energy E' given by

$$E' \sim \gamma^2 E,$$

where E is the original energy of the photon and γ is the relativistic factor for the electron $(1/\sqrt{(1 - v^2/c^2)}$; cf. p. 23). Consider now the electrons that are responsible for the galactic radio background through their synchrotron emission. A typical energy for such an electron might be, say, 1 BeV. Its γ would then be 2000 and with $E \sim 10^{-3}$ eV, we see that the scattered photon would be raised in energy to about 4 kilovolts. This takes us right into the X-ray region at a wavelength of 5 ångströms. The Galaxy would thus be an X-ray source. We must ensure, of course, that its strength is compatible with the observed X-ray background. Now it turns out that the inverse Compton effect and the synchrotron mechanism are essentially the same process of a relativistic electron emitting radiation while under the influence of an electromagnetic field. Consequently they lead to a similar rate of energy transfer from the electrons if the energy density in the photon field and the magnetic field are comparable. This is just the case for our Galaxy, as we have seen. Thus the known energy flux in the galactic radio background should be of the same order as the energy flux in the emitted X-rays. This flux would be about 1 per cent of the observed X-ray background.

The proposed black body radiation field thus passes this test. The margin by which it does so is not as large as it might seem, however, because the rate of energy transfer to the X-rays is proportional to the Compton collision rate and so to the photon density in the black body radiation field, that is, to the third power of its temperature. Moreover if the radiation field had a higher temperature, the mean energy of its photons would be greater. Accordingly an electron of lower energy would suffice to produce a given X-ray energy. We must therefore allow for the fact that in our Galaxy there are more electrons at the lower energies. The net result of all this is that if the black body background had a temperature of only 10 °K, the X-ray flux from the Galaxy would be of the same order as the observed X-ray flux. This would be inconsistent with the observed fact that this X-ray flux is isotropic to a precision of about 10 per cent.

There is another important aspect of this problem and that is the resulting energy drain on the electrons. For $T \sim 3$ °K, the drain due to the production of X-rays is of the same order as that due to the radio emission (or to Compton collisions with starlight photons which also have an energy density of about 1 eV cm^{-3}). If we tried to push T up, however, this drain would increase very rapidly, roughly like T^4. It would soon become so high that it would be very difficult to understand where the electrons got their energy from in the first place. For $T \sim 10$ °K, for instance, a 1 BeV electron would lose half its energy in 10 million years, instead of the billion years characteristic of the synchrotron process alone. A further point is that the observed energy spectrum of the electrons (fig. 50) shows no sign of significant energy losses out to at least 50 BeV, at which energy the lifetime of an electron in the 3 °K radiation field is only about 20 million years. This tells us that these electrons must leak out of the local trapping region in a shorter time than this. In a 10 °K radiation field the leakage time would have to be less than 3×10^5 years. This would have very severe implications for the propagation of cosmic ray protons and heavier particles.

These considerations take on even greater importance when we apply them to a radio source with a substantial red shift z. The temperature of the black body radiation at such a source would be increased by the factor $1+z$ over its local value, and the all-

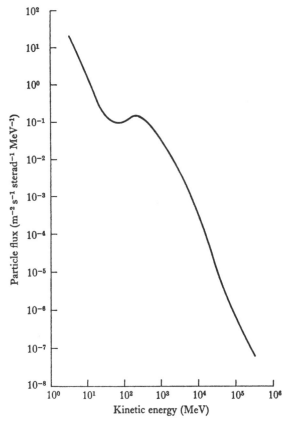

Fig. 50. The energy spectrum of cosmic ray electrons. Note that there is no sign of a significant deviation from a simple power law out to at least 50 BeV.

important energy density by $(1+z)^4$. The balance between the synchrotron and Compton effects that holds in our Galaxy would then be disturbed, unless the magnetic fields in radio galaxies are much greater than in our Galaxy. In any case the Compton lifetime of a relativistic electron of given energy would be at least 81 times shorter in a radio source with a red shift of 2 than in our own Galaxy. This effect may contribute significantly to the evolution of radio sources which is so important for analysing the source counts. The situation would be even more extreme for a radio source with a red shift of, say, 4. At such a red shift the life-time of an electron interacting with the black body radiation would be reduced by a factor 625 as compared with a contemporary electron. These interactions would also give rise to X-rays, which might make the dominant contribution to the observed X-ray background, especially if one allows in addition for the evolutionary increase in the power or number of radio sources with large red shifts.

It is also interesting to consider the Compton collisions that might be occurring in intergalactic space. To estimate the resulting X-ray flux we observe that the path length along a line of sight through the Universe is about 10^5 times that through the Galaxy. Since the galactic X-ray flux is expected to be about one per cent of the observed background, an intergalactic flux of relativistic electrons which is 10^{-3} of the galactic flux, would suffice to account for the observed X-ray background. This looks an attractive explanation at first sight because, as we saw in chapter 10, the intergalactic flux of cosmic ray protons that have leaked out of galaxies may well be 10^{-3} or even 10^{-2} of the galactic flux. However, we must not overlook the losses arising from the inverse Compton process itself. These losses lead to a lifetime for the electrons of only about a billion years, one tenth of the expansion time-scale. As a result the intergalactic flux of electrons would be far too small to account for the observed X-ray background. The only hope would be if electrons leaking from radio galaxies and QSOs led to a substantially larger flux. There are no doubt more relativistic electrons in such objects than in a typical galaxy but there are far fewer such objects, and we do not know enough to work out this balance at all convincingly. It therefore remains possible but unproven that the X-ray background is due to inverse Compton collisions between intergalactic relativistic electrons and the black body radiation. At the time of writing it seems more likely that the X-ray background arises mainly from inverse Compton collisions in radio sources with red shifts in the range from 3 to 5.

COSMIC RAYS AND THE MICROWAVE RADIATION FIELD

From the point of view of a cosmic ray proton of 10^{20} eV, which has a γ of 10^{11}, a photon of 10^{-3} eV looks like one of $10^{-3} \gamma$ eV or 100 MeV. Such an energetic photon striking a stationary proton

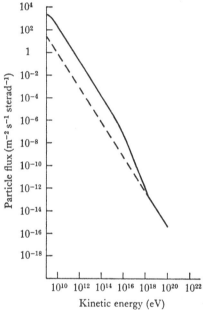

Fig. 51. The energy spectrum of cosmic rays. Note the slight flattening of the spectrum beyond 10^{18} eV and the absence of a sharp cut-off close to 10^{20} eV.

would be essentially at the threshold for producing a pion (rest-mass $\sim$ 137 MeV). This means that from the terrestrial point of view a cosmic ray proton of 10^{20} eV could collide with a microwave photon, produce a pion, and so be degraded in energy. Now cosmic rays with energies in the range 10^{18}–10^{20} eV are sometimes thought to have an extragalactic origin because they cannot be confined by the magnetic field of the Galaxy (chapter 2), and so their flux near the Earth cannot be built up for the storage time of a few million years that is characteristic of lower energy cosmic rays. Thus if all the cosmic rays originated in the Galaxy we might expect a sharp drop in the flux at about 10^{18} eV. This sharp drop is not found (fig. 51) and indeed the spectrum appears to *flatten* slightly in this region.* This suggests, although it does not prove, that there exists an intergalactic flux of cosmic rays with the flatter spectrum that dominates the galactic flux locally at energies above 10^{18} eV.

Now the observed spectrum shows no signs of a cut-off right out to cosmic ray energies of 10^{20} eV, which is the maximum energy that existing techniques can detect. This limit is tantalisingly close to the threshold for severe attenuation by the microwave background. Fortunately experiments are now being conducted that should enable the spectrum to be extended to the range 10^{21}–10^{22} eV. If there are still no signs of attenuation found, and if the microwave radiation really fills intergalactic space, it would mean that these ultra-high-energy cosmic rays have a lifetime less than 100 million years, and so could have come from no further than 100 million light years (the Virgo cluster of galaxies is 30 million light years away). On the other hand if severe attenuation is found it would strongly suggest that these cosmic rays come from still greater distances.

γ-RAYS AND THE MICROWAVE RADIATION FIELD

In this section we shall see that high energy cosmic γ-rays (if there are any) would also be degraded by interaction with microwave photons. We cannot now look at the collision from the rest-frame of the high energy object, since we cannot bring the γ-ray to rest, but what we can do instead is to use a frame of reference in which the γ-ray and the microwave photon have the same energy. If the energy of the γ-ray is E relative to the Earth, and we transform to a frame moving with velocity v, then the energy of the γ-ray becomes E/γ $(\gamma = 1/\sqrt{(1 - v^2/c^2)})$ and the energy of a microwave photon becomes $10^{-3}\gamma$ eV. Setting these equal we have

$$E/\gamma = 10^{-3}\gamma.$$

* Recent cosmic ray data suggest that there is no flattening of the spectrum beyond 10^{18} eV.

Let us now choose E so that the common energy of each photon is just the rest-energy of an electron. We would then be at the threshold for electron–positron pair creation. This requires that

$$\frac{E}{\gamma} = 10^{-3}\gamma = m_0 c^2.$$

Eliminating γ we have $10^{-3}E = (m_0 c^2)^2$

or $E = 2.5 \times 10^{14}$ eV.

At this energy and above, cosmic γ-rays would be degraded by the microwave photons through pair creation.

The effect is a large one. The cross-section σ for the process near threshold is roughly the square of the classical electron radius, $\sim 10^{-25}$ cm^2, and so the mean free path which would be about $1/n\sigma$, where n is the density of the microwave photons, becomes $\sim 10^{22}$ cm. This is smaller than the size of the Galaxy. There should therefore be a sharp cut-off in the spectrum of the cosmic γ-ray background at about 2.5×10^{14} eV. This cut-off should continue right out to 10^{21} eV. Similarly, discrete γ-ray sources would be subject to this strong absorption.

Unfortunately no γ-rays above an energy of 100 MeV have yet been reliably detected from cosmical objects or in the form of a diffuse background. The energy range above 10^{14} eV is under experimental study at the moment but it is not clear whether the actual flux is large enough to be detectable. It seems likely that significant results will come first from the proton measurements.

CONCLUSIONS

The phenomena described in this chapter show in a striking way one of the most fascinating features of modern astrophysics, namely the close relationship that exists between processes involving totally different energy or time-scales. For instance, measurements of the background flux of 10^{-5} eV photons (the microwave measurements), in conjunction with considerations from cosmology and nuclear physics, enable one to discuss processes in which 10^{-3} eV photons are converted into 10^3 eV photons by 10^9 eV electrons, or are able to absorb or degrade 10^{14} eV photons and 10^{22} eV protons. One is left wondering what new relationships remain to be discovered.

16 The isotropy of the cosmic microwave radiation

INTRODUCTION

When Penzias and Wilson first discovered the excess microwave background radiation they found that its intensity was much the same in all directions. The precision to which this isotropy was established was initially given as 10 per cent, but a later study of their records enabled them to reduce the uncertainty to 3 per cent. This high degree of isotropy made it from the beginning unlikely that the excess radiation arises in the Galaxy with its strongly asymmetric shape. The Universe, on the other hand, was considered to be fairly isotropic on a large scale, and this lent support to the idea that the origin of the radiation is cosmological. The Universe is not exactly isotropic, however, and the question arises of the possible angular scale and amplitude of intensity variations in the background radiation. Recent observational studies of this question have attained a precision much greater than 3 per cent, and represent the most accurate measurements yet made in cosmology. They also provide us with powerful information of a completely new character concerning the structure of the Universe on various scales. It is these new questions that we wish to discuss in this final chapter.

In order to appreciate the significance of the angular distribution of the background radiation it is helpful to remember that when the radiation we now observe was in regions that have a large red shift relative to us it would have interacted strongly with free electrons through the Thomson effect (p. 144). The radiation ceased to be effectively scattered at a red shift z_0 that depends upon the cosmological model and the thermal history of the intergalactic gas, but in no case is this red shift less than about 7 (the value that corresponds to one mean free path in a completely ionised high density Universe). From our present point of view we may regard the electrons that last effectively scattered the radiation as being the sources of the radiation, just as the photosphere of the Sun is the effective source of solar radiation. Moreover we cannot 'see' further into the Universe than a red shift z_0, just as we cannot see the interior of the Sun* – all the information is smoothed out by scattering until the scattering itself stops. Thus the degree of isotropy of the background radiation tells us something about the surface of last scattering and our relation to that surface. In order to understand what this information is we must consider the various effects that can introduce anisotropy into the radiation.

SOURCES OF ANISOTROPY IN THE BACKGROUND RADIATION

The most obvious source of anisotropy is the peculiar velocity of the observer. Even if the radiation is otherwise isotropic it would not appear to be so to an observer moving through it. By virtue of the Doppler effect such an observer would see an increased intensity in front of him, a decreased intensity behind him, and a characteristic (cosine) dependence at intermediate angles. *The isotropic radiation field thus defines a rest frame.* Any observer moving relative to this frame could in principle measure his velocity by studying the angular distribution of the radiation. As the sky is scanned from the rotating earth the intensity of the background would have a 24-hour period in any plane, with an amplitude depending on his velocity and on the angle his velocity makes with the plane.

It is also possible to have anisotropies inherent in the radiation that are not removable by any choice of the observer's motion. If, for instance, the Universe expands at a different rate in different directions, then once the radiation no longer interacts with matter it would cool off at different rates in these different directions. If

* This refers to electromagnetic radiation only. Neutrinos, with their extremely low scattering cross-section, come to us directly from the centre of the sun, and also from red shifts much larger than z_0.

one thinks in terms of the expanding box whose walls are perfect mirrors (p. 157) this becomes immediately evident. If the different pairs of opposite faces have different velocities then the radiation would suffer different red shifts when it is reflected from the different pairs of faces. If the Universe is still taken as homogeneous then the faces of an opposite pair would have equal and opposite velocities. In this case the intensity of the radiation in opposite directions would be the same and the variations in any plane would have a 12-hour period. We must also allow for the effects resulting from the scattering of the radiation by a re-heated intergalactic gas in an anisotropic universe. Rees has shown that this could have a measurable effect on the spectrum (and the polarisation) of the radiation even for an anisotropy that would be barely detectable directly.

Finally we consider the more complicated case when the Universe is inhomogeneous. There are now various possibilities. First of all the surface of last scattering may be inhomogeneous. The radiation intensity itself could vary over the surface, or the velocity of the electrons in it could be variable, giving rise to a Doppler shift that would depend on the direction of observation. Secondly, there could be density inhomogeneities between the scattering surface and ourselves. The effect of such inhomogeneities on the intensity of the radiation passing through them is quite complicated. We shall return to this question later; here we merely note that there would be a gravitational red shift effect.

We may summarise this discussion by saying that there are three basic sources of anisotropy in the background radiation: (i) the peculiar velocity of the observer, giving a 24-hour period; (ii) an anisotropy in the expansion of the (homogeneous) Universe, giving a 12-hour period; (iii) inhomogeneities in the Universe, giving an irregular angular distribution. We now consider the observations relating to these types of angular distribution.

OBSERVATIONAL LIMITS ON THE ANISOTROPY IN THE BACKGROUND RADIATION

(i) The 24 hour period. A search for anisotropy with a 24-hour period in a plane close to the celestial equator has been made by Partridge and Wilkinson, by Conklin and by Henry. Measurements of this type can attain a higher precision than an absolute measurement of the background temperature because one is here concerned with an intercomparison of different regions of the sky, and calibration problems are less important. Partridge and Wilkinson found no anisotropy with a precision of 0.03 ± 0.07 per cent (of 3 °K) (fig. 52). The corresponding restriction on the peculiar velocity of the observer in the plane of observations is about 0.1 per cent of

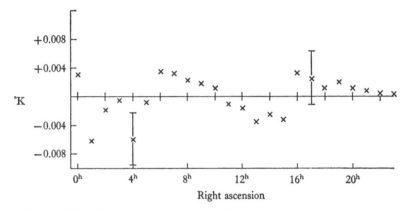

Fig. 52. The distribution of background temperature along the celestial equator, according to Partridge and Wilkinson. More than two years of data, obtained at Princeton and Yuma, have been averaged. The fluctuations about the mean are not regarded as significant. (After R. B. Partridge, *American Scientist* **57**, 37 (1969).)

the speed of light, or 300 km s^{-1}. The significance of this important result will be discussed below.*

(ii) The 12-hour period. Partridge and Wilkinson also searched for a 12-hour period in their observations. Again nothing significant was observed.

(iii) Small scale irregularities. These have been looked for by Partridge and Wilkinson and by Conklin and Bracewell. Partridge and Wilkinson failed to find local hot or cold spots with a precision of 0.5 per cent. Conklin and Bracewell also found no evidence of small scale irregularities (fig. 53). They studied carefully a small strip of sky and were able to place a limit on

* Conklin and Henry both reported a positive result with marginal significance, but this has not yet been generally accepted.

fluctuations in the background of less than 0.2 per cent on an angular scale of about 1 degree. In later work they were able to reduce this limit to 0.05 per cent. By a special technique they also obtained limits on a scale of less than the resolving power of their receiver. On a scale of θ minutes of arc their limit was $2/\theta$ per cent for $0 < \theta < 10$, and $2/3\theta^{\frac{1}{2}}$ per cent for $10 < \theta < 120$. It may not be possible to improve on this limit because of fluctuations arising from discrete radio sources. Even so the limit already places severe restrictions on possible irregularities in the surface of last scattering.

THE PECULIAR VELOCITY OF THE SUN

We have seen that the absence of a detectable 24-hour variation in the background limits the velocity of the sun in directions close to

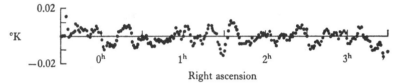

Right ascension

Fig. 53. The distribution of background temperature along the celestial equator, according to Conklin and Bracewell.

the celestial equator to about 300 km s^{-1}. This result is of great significance. In order to understand it we must first answer the question: relative to what is this limiting velocity measured? It is in the first place relative to the radiation field itself, which is able to define a rest-frame. However the immediate sources of this field are the electrons on the surface of last effective scattering. Of course these last scatterers do not lie on a surface of no depth, any more than the visible surface of the Sun is a strict surface. However, for many purposes it suffices to think of the material one mean free path away as the effective source, and this material has a red shift of at least 7 (p. 194). We can thus regard the limiting velocity of 300 km s^{-1} as being relative to very distant matter in the Universe.

This situation reminds us of Mach's principle which was briefly referred to on p. 125, and is discussed in detail in my book *The Physical Foundations of General Relativity*. According to this principle local inertial frames are those unaccelerated relative to the bulk of the matter in the Universe. In particular, non-rotating frames, that is, those that do not involve centrifugal or Coriolis forces, must be those relative to which distant matter can be seen to be non-rotating. If this principle is correct it is of profound theoretical importance. Our immediate concern, however, is the precision to which it can be established that the two forms of non-rotation are in agreement. A familiar example of this agreement is that the stars go round the Earth once a day and that a Foucault pendulum at one of the Earth's poles would also go round once a day. However, a Foucault pendulum is not a very accurate instrument for determining a non-rotating frame. Modern gyroscopes are better but are still not very precise. Considerable improvements may be expected from the use of superconducting gyroscopes, but at the moment the most accurate non-rotating frame is determined by the rotation of the Galaxy. The Galaxy rotates with an angular velocity which at the Sun is about 0.5 seconds of arc per century, and the resulting centrifugal forces in the rotating frame of the Galaxy are responsible for its flattened shape. Thus this angular velocity is a velocity relative to a dynamically non-rotating frame. Now the rotation of the Galaxy can be detected not only dynamically from its flattened shape but also kinematically by observing the motions of stars so far out from the centre that their angular velocity is much less than that of the Sun (we here exploit the *differential* rotation of the Galaxy, discussed on p. 19). Preliminary attempts are also being made to determine the rotation of the Galaxy using galaxies and QSOs rather than distant stars. The dynamical and kinematical methods are in rough agreement, which thus confirms the coincidence of the two definitions of non-rotation to a precision of about 0.5 seconds of arc per century.

Now as a result of its rotation in the Galaxy the sun has a velocity of about 250 km s^{-1}. According to our present discussion this velocity should be a velocity relative to distant matter and so to the background radiation field. Thus, as Partridge and Wilkinson have pointed out, a slight improvement in the precision with which 24-hour variations in the intensity of the background can be measured should lead to a positive result, and one which would provide confirmation of this whole structure of ideas.

However, there are some complications. The Galaxy as a whole could have a linear velocity through the background radiation field as well as a rotational velocity. This indeed is believed to be the case. An analysis of the velocities of galaxies in the Local Group indicates that our Galaxy has a velocity relative to the Local Group as a whole of about 100 km s^{-1}. In addition there is some evidence that the Local Group belongs to a supercluster of galaxies whose centre lies in the Virgo cluster. This supercluster also appears to be flattened and it has been suggested that this is a sign that the supercluster as a whole may be rotating. This possibility has recently been reanalysed using modern data on distances and velocities of nearby galaxies. The results are quite uncertain, but it does seem that the Local Group may have a rotational velocity of about 300 km s^{-1} around the Virgo cluster, thus complicating the comparison of any future observed 24-hour variations of the background with the known rotation of the Galaxy. In addition one may have to allow for differential expansion of the supercluster and even for a motion of its centre of gravity relative to distant matter.

If these complications can be sorted out, and if the supercluster is found to be rotating at the expected rate, it would become possible to test Mach's principle with much greater precision. The reason is that although the velocity of the Local Group around Virgo would be of the same order as the velocity of the Sun around the centre of the Galaxy, the *angular* velocity of the Local Group would be much less than that of the Sun in the Galaxy, because the Virgo cluster is much further away than is the centre of the Galaxy. In fact, if the rotation of the supercluster is detected by future measurements of 24-hour variations in the background, it would test Mach's principle with a precision of about 10^{-3} seconds of arc per century, a 500-fold increase over the present accuracy of the comparison. Thus the future measurements of the background may enable us to derive important information concerning both the nature of the local supercluster and the validity of Mach's principle.*

* Hawking has recently shown that the observed isotropy of the microwave background limits the 'absolute' angular velocity of the Universe to less than about 10^{-6} seconds of arc per century.

THE ISOTROPY OF THE UNIVERSE

The limits placed by Partridge and Wilkinson on the 12-hour variation of the background ensure that the Universe is expanding isotropically to considerable precision. The question then arises: why is the Universe so isotropic? One possible answer would be that the singularity out of which the Universe emerged was itself isotropic; in other words that it is a matter of initial conditions. This answer does not seem very satisfactory and a better one has been proposed by C. W. Misner and by A. G. Doroshkevich, Y. B. Zel'dovich and I. D. Novikov. According to this proposal the Universe may initially have been highly anisotropic, but physical processes occurring during the early stages of the expansion would have reduced the anisotropy to a very small value. From Misner's point of view this would be one example of a general proposition, namely that a large part of the present structure of the Universe may be more or less independent of the initial conditions. If this proposition is correct one should be able to 'predict' many of the observable features of the Universe unambiguously from general relativity and the other relevant laws of physics. The study of what we may call Misner's programme is only in its infancy and it seems likely to be of growing importance in the theoretical cosmology of the future.

Meanwhile we would like to give a brief account of Misner's work on the dissipation of anisotropy in the expanding Universe. He initially restricted himself to homogeneous models in which space at any one time is flat (analogous to the case $k = 0$ of the Robertson–Walker models). The most important dissipation mechanism appears to arise from the interaction between thermally excited neutrinos and electron–positron pairs when the Universe had a temperature of about 10^{10} °K. At that time the neutrinos would suffer different Doppler shifts in different directions if there was an anisotropy in the expansion rate of the Universe. However, the collisions between the neutrinos and the electron pairs would tend to isotropise the neutrinos and this in turn would tend to isotropise the expansion, since the dynamics of the Universe would be strongly influenced by the gravitational effects of the neutrinos. The importance of this neutrino viscosity depends on the strength of the interaction between neutrinos and electron-pairs. This interaction-strength has not yet been measured in the laboratory but it is believed to be governed by a theory that gives a very good account of similar interactions whose strengths are known experimentally. This Feynman–Gell-Mann theory has a built-in symmetry that relates the neutrino–electron interaction to the known interactions. Misner has shown that if this theoretical interaction strength is correct the 12-hour anisotropy in the background radiation would now be less than 0.03 per cent for a large range of initial anisotropies. This result is consistent with the observations of Partridge and Wilkinson which place an upper limit of about 0.2 per cent on the 12-hour anisotropy. Misner also showed that for the same range of initial anisotropies any anisotropy surviving at temperatures of about 10^9 °K would be too small to influence the primordial formation of helium (cf. p. 174). These results are clearly of great importance and it is to be hoped that they can be carried through in the more general models that have yet to be treated.*

THE INHOMOGENEITY OF THE UNIVERSE

The small scale inhomogeneities in the Universe (galaxies and galaxy clusters) have a negligible influence on the microwave radiation. However we saw in chapter 7 that the QSOs might possibly be clustered on a scale of about 1000 Mpc, and that this clustering might be related to large scale density fluctuations in the material of the Universe. Such a large scale density fluctuation could have a measurable influence on the microwave background, an idea first proposed by R. K. Sachs and A. M. Wolfe. If one calculates the intensity profile across a region of high density surrounded by a region of low density, one obtains the results illustrated in fig. 54. A region 750 Mpc across with a density contrast of three to one at a red shift of 1.5 would produce an intensity fluctuation of about 0.3 per cent on an angular scale of about 15 degrees. Such a fluctuation should be detectable in the future.

* Recent work suggests that Misner's programme of accounting for the observed symmetry of the Universe in terms of dissipative processes may have significant limitations.

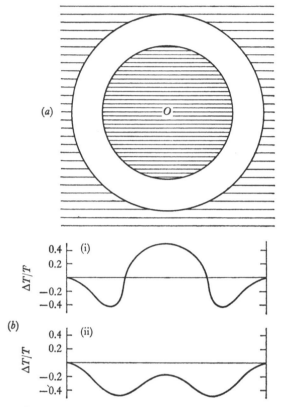

Fig. 54. (*a*) A spherically symmetric perturbation in the Universe.
    (*b*) Temperature profiles across such a perturbation with ~ 3 times the background density. $\triangle T/T$ is expressed in units of GMV/c^3R, where V and R are the velocity and radius of the unperturbed boundary. The profiles are drawn for (i) an undecelerated background universe and (ii) an Einstein–de Sitter universe.
(From M. J. Rees and D. W. Sciama, *Nature* **217**, 511 (1968).)

CONCLUSIONS

It should be evident to the reader that throughout this chapter we have been unable to press home our arguments because the observations are not quite adequate to support them. What is significant, however, and what has justified the writing of this chapter, is that the observations are probably only at their beginning. There may be many more QSOs of large red shift waiting to be discovered, while the study of the angular distribution of the cosmic microwave radiation is only in its infancy. Eventually the whole sky should be mapped with a relative precision of 0.1 per cent on a variety of angular scales. When that has been done we should possess powerful new information on the validity of Mach's principle, on the existence and structure of the local supercluster of galaxies, and on the properties of large scale fluctuations in the distribution of matter. Here surely we stand at the threshold of the cosmology of the future.

Further reading

More advanced treatments of modern cosmology may be found in
   M. S. Longair: Observational cosmology, *Reports on Progress in Physics*
   34, 1125 (1971).
P. J. Peebles: *Physical Cosmology* (Princeton 1971).
S. Weinberg: *Gravitation and Cosmology* (Wiley, New York 1972).
S. W. Hawking and G. F. R. Ellis: *The Large Scale Structure of Space–*
   Time (Cambridge 1973).
Y. B. Zel'dovich and I. D. Novikov: *Relativistic Astrophysics*, vol. 2
   Cosmology (Chicago 1975).
Proceedings of the International School of Physics Enrico Fermi: Course
   47, General relativity and cosmology (ed. R. K. Sachs: Academic
   Press, New York and London 1971).
Lecture Notes from Institut d'Etudes Scientifique de Cargese, July 1971
   (ed. E. Schatzman: Gordon and Breach, New York).
E. R. Harrison: Standard model of the early universe, *Annual Review*
   of Astronomy and Astrophysics **11**, 155 (1973).
I. D. Novikov and Y. B. Zel'dovich: Physical processes near cosmo-
   logical singularities, *Annual Review of Astronomy and Astrophysics*
   11, 387 (1973).
IAU Symposium No. 63. *Confrontation of Cosmological Theories with*
   Observational Data (ed. M. S. Longair, Reidel, Dordrecht-Holland
   1974).

The older work on cosmology may be traced through the following books
which contain extensive references:
   H. Bondi: *Cosmology* (Cambridge, second edition 1960).
   J. D. North: *The Measure of the Universe* (Oxford 1965).

Astronomical topics of importance to cosmology are discussed in:
   P. W. Hodge: *Physics and Astronomy of Galaxies and Cosmology*
      (McGraw-Hill, New York 1966).
   G. R. and E. M. Burbidge: *Quasi-stellar Objects* (Freeman, San
      Francisco and London 1967).
   M. Schmidt: Quasars, *Annual Review of Astronomy and Astrophysics* **7**,
      527 (1969).
   M. Ryle: The counts of radio sources, *Annual Review of Astronomy*
      and Astrophysics **6**, 249 (1968).

R. J. Tayler: *The Origin of the Chemical Elements* (Wykeham, London 1972).

G. B. Field: Intergalactic matter, *Annual Review of Astronomy and Astrophysics* **10**, 227 (1972).

P. Thaddeus: The short-wavelength spectrum of the microwave background, *Annual Review of Astronomy and Astrophysics* **10**, 305 (1972).

W. D. Arnett: Explosive nucleosynthesis in stars, *Annual Review of Astronomy and Astrophysics* **11**, 73 (1973).

J. Silk: Diffuse X- and γ-radiation, *Annual Review of Astronomy and Astrophysics* **11**, 269 (1973).

D. N. Schramm: Nucleo-cosmochronology, *Annual Review of Astronomy and Astrophysics* **12**, 383 (1974).

B. Zuckerman and P. Palmer: Radio radiation from interstellar molecules, *Annual Review of Astronomy and Astrophysics* **12**, 279 (1974).

H. Reeves: On the origin of the light elements, *Annual Review of Astronomy and Astrophysics* **12**, 437 (1974).

IAU Symposium No. 58. *The Formation and Dynamics of Galaxies* (ed. J. R. Shakeshaft, Reidel, Dordrecht-Holland 1974).

Lively brief articles on current developments may be found in the following journal: *Comments on Astrophysics and Space Physics*.

Author Index

Subject Index

absorption lines
  in optical spectra of QSOs, 77, 138
  produced by clusters of galaxies too
    faint to be seen, 137
absorption troughs, 130, 131, 134
alpha particles, in cosmic rays, 25, 151
ammonia, in interstellar gas, 21
Andromeda galaxy, 39, plate 4
  velocity of, 42, 43
  weak radio source, 54, 56, pla.e 11
angular diameter
  relation of red shift, luminosity-
    distance, and, 122–3
  of Sun, 1
anti-neutrinos, 167, 173, 174
atmosphere
  absorption by, 136, 176, 179, 181
  microwave emission from, 178
atomic quantities, numerical coinci-
    dences between cosmological
    quantities and, 100, 124–5, 144
beryllium, in cosmic rays, 26
'black body' cosmic radiation, 141,
    157, 159–60, 167, 168, 185–92
  detection of, 176–84
  isotropy of, 193–201
  temperature of, 130, 140, 166, 169,
    170, 178, 179
  See also microwave background,
    cosmic
blue dwarf stars, 2
blue shifts, absent among QSOs, 81–2
blue stellar objects, 62, 72
boron, in cosmic rays, 26
carbon
  conversion of helium to, 8
  in cosmic rays, 151
  in intergalactic gas, 136, 138
  in QSOs, 70
Cassiopeia radio source, 53

Centaurus A radio source, 50, plate 9
  multiple, 54, 58
  optically identified, 54, 55
  repeated explosions of? 61
Cepheid variables
  type I in external galaxies, 39, 41, 42
  type II in Galaxy, 16–18, 41, 42
Chandrasekhar limit, 9
clusters
  of galaxies, 97, 126, 129, 199, 203,
    plate 8; 'local', 39, 40, 42
  globular, of stars, 2, 18–19, 37
  of QSOs, 94–7
colour
  of QSOs, 63–5
  and surface temperature, 2
Compton collisions between relativistic
    electrons and photons, 77, 147–8,
    186, 187, 188, 189
convection, in stars, 154
cosmic rays, 22–6, 147, 187
  accelerated in pulsars? 30
  electrons of, 26–7, 147–8, 188
  Galactic magnetic field and, 36; in-
    tergalactic magnetic field and, 148
  heating of intergalactic gas by, 140–
    1, 147
  microwave radiation and, 190–1
  solar component of, 150
cosmical constant, 99, 113
cosmological quantities, numerical
    coincidences between atomic
    quantities and, 100, 124–5, 144
Crab nebula, 8, plates 1, 6, 9
  pulsar in, 11, 30
  radiations from, 30, 31, 54
cyanogen, interstellar, 166, 176, 181–4
Cygnus A radio source, 50, 53, 55, 57,
    plate 12
  optically identified, 50, 54–5